INSECT–PLANT INTERACTIONS AND INDUCED PLANT DEFENCE

Chairman's introduction

John A. Pickett

Biological and Ecological Chemistry Department, IACR-Rothamsted, Harpenden, Hertfordshire A L5 2JQ, UK

We in the 'developed' countries are probably naïve in believing that we will continue to have a surplus of agricultural products for eternity. For the developing countries, the reality is one of food shortages. Although famines are fairly uncommon, there exist tremendously deprived areas in terms of the kind of food that people want to eat. With the world population set to rise to 8.5 billion by 2020, we are going to have to put a lot more technology into food production: there simply isn't a large enough area of land available for cultivation. If we don't use technology to improve agricultural production, we will stress even more the unexploited natural habitats that remain.

In introducing this subject, I'd like briefly to mention, by way of an example, some work we have been doing in Kenya. Here, as in many other parts of Africa, there can be tremendous crop losses through damage by lepidopterous larvae, particularly stem-borers, and the aggravated wind lodging of maize and sorghum crops. The farmers can't address pest problems by using pesticides simply because they can't afford them: even though Kenya's economy is quite a thriving one, it doesn't spend a lot of hard currency buying in basic agricultural materials such as fertilizers and pesticides: if it did, it would start to find itself in debt or without essential pharmaceuticals. However, there are alternatives. In collaboration with ICIPE in Kenya, we are developing an intercropping system which exploits signals sent out from plants to insects, telling them whether or not the plant from which the chemicals come is a suitable host (Pickett et al 1997, Hick et al 1997). By creating such a 'push–pull' situation within the maize crop, it is possible to control these stem-borers. There's also some evidence in this work of plants actually signalling between themselves. This is an aspect which has attracted media attention, and it's something that we would be able to capitalize on if we were able to characterize the kind of signals that are induced during plant stress. This is an exciting area, and one of which I feel we are only at the beginning. We all have to obtain resources and speak to the media: in doing this, we obviously put the best complexion on what we're doing. However, at this symposium, we need to take an objective appraisal of exactly where we are, and where we perhaps should be going in the future.

When I was asked to put together this meeting, I felt I needed some advice from a distinguished biologist. I would like to thank Marcel Dicke, who fulfilled this role very ably. With these thoughts in mind, I would like to start the meeting.

References

Hick AJ, Pickett JA, Smiley DWM, Wadhams LJ, Woodcock CM 1997 Higher plants as a clean source of semiochemicals and genes for their biotechnological production. In: Wrigley S, Hayes M, Thomas R, Chrystal E (eds) Phytochemical diversity: a source of new industrial products. Royal Society of Chemistry, Cambridge (Special Publication 200) p 221–236

Pickett JA, Wadhams LJ, Woodcock CM 1997 Developing sustainable pest control from chemical ecology. Agric Ecosyst Environ 64:149–156

Evolutionary aspects of plant–carnivore interactions

Louise E. M. Vet

Laboratory of Entomology, Wageningen Agricultural University, P.O. Box 8031, NL-6700 EH Wageningen, The Netherlands

Abstract. Plants can respond actively to damage by herbivores. In addition to a mode of defence that is directly aimed at the herbivore itself, plants can emit volatiles that attract carnivores, i.e. the enemies of their enemies. Knowledge of the mechanisms underlying the induction of these herbivore-induced plant volatiles and of the responses of the carnivores is progressing rapidly. Inferences on the initial causes of evolution of herbivore-induced plant volatiles remain conjectural. However, once plant–carnivore interactions have evolved to the net benefit of both participants this mutualism is expected to have evolutionary and ecological consequences for the three trophic levels involved. When plant selection and foraging behaviour of natural enemies is linked to plant fitness this can influence different aspects of the plant defence strategy. The way carnivores perceive and process plant information may influence the evolution of the plant signal (i.e. quantitative and qualitative composition of the odour blend in response to herbivore damage). Vice versa, the signal-to-noise ratio of the information may influence the way carnivores respond to plant cues (innately or through learning). Selection will act on herbivores to disconnect the plant–carnivore link, for example by boycotting the informational value of herbivore-induced synomones. Through plant selection and feeding behaviour herbivores can influence their chance of being found by carnivores. Hence, responses of carnivores to plant cues can influence the evolution of food-plant use by herbivores. The conspiracy between plants and carnivores is at the heart of evolutionary ecology, and wide open for experimental and theoretical investigations.

1999 Insect–plant interactions and induced plant defence. Wiley, Chichester (Novartis Foundation Symposium 223) p 3–20

In spite of the fact that half of the insect world munches plants, the world is still green. Why that is so has been a matter of debate between ecologists, some favouring that it is mainly the defensive ability of plants (bottom-up effect), others emphasizing the suppressive power of carnivores that attack the herbivorous arthropods (top-down effect) (Strong et al 1984). As often, the truth is probably somewhere in the middle, whereby both bottom-up and top-down effects reduce the impact of herbivores on plants, acting in concert but varying in their degree of importance (Hunter & Price 1992).

Plants can certainly defend themselves very effectively. Well known is the strategy of *direct* defence that is aimed at the herbivore directly. Direct defence can be mediated by plant chemicals such as toxins, repellents or digestibility reducers, or by plant structures such as thorns and spines. But even when herbivores can break through this first line of defence, plants are not left helpless victims. Plants can additionally defend themselves by enhancing the effectiveness of natural enemies of herbivores, so-called *indirect* defence (Price et al 1980). Indirect defence can involve the provision of shelter for carnivores (e.g. domatia) or the provision of alternative food such as floral and extrafloral nectar. With a few exceptions, these strategies are forms of constitutive indirect defence, acting before herbivore attack and aimed at preventing this occurring. When herbivores do attack, plants have a strategy of inducible indirect defence: they can emit volatiles that attract and arrest carnivores. This intriguing and seemingly mutualistic conspiracy between plants and carnivores has aroused the interest of many ecologists and the subject of plant–carnivore chemical communication is presently a fast growing research area in chemical ecology.

Plants can provide the enemies of their enemies with chemical information about the presence, identity and densities of herbivores. The benefit for carnivores in responding to these plant cues is high, since the possibility of using herbivore-derived information is limited because of strong selection on herbivores not to reveal themselves (Vet et al 1991, Vet & Dicke 1992).

A wealth of empirical data has been produced on this mode of indirect defence of plants. (For reviews I refer to Vet & Dicke 1992, Tumlinson et al 1993, Turlings et al 1993, Dicke 1994, Dicke & Vet 1999). Different approaches, often successfully combined, are being applied to unravel the mysteries of plant–carnivore interactions. First the *mechanistic approach*, focusing on the 'how' of the interactions. Studies include the identification of the plant chemicals involved (e.g. Turlings et al 1990), the biosynthetic pathways that underlie volatile production (e.g. Hopke et al 1994), and the identification of the herbivore factors that initiate the induction process (e.g. Mattiacci et al 1995, Alborn et al 1997). On the side of the responding carnivore, research is aimed at identifying exactly which cues carnivores respond to behaviourally and which behaviours are evoked by each of the cues (e.g. Geervliet et al 1996, Powell et al 1998). In the present contribution I will focus on a second approach, the *functional approach*, where we study the 'why' (i.e. the evolutionary aspects) of plant–carnivore interactions. Here the sorts of questions that are being asked include the following. How did plant–carnivore mutualism evolve and under which conditions is signalling an 'evolutionarily stable strategy' (ESS, after Maynard Smith 1982, e.g. Godfray 1995, Sabelis & De Jong 1988)? Why do carnivores need plants (e.g. Vet et al 1991, Vet & Dicke

1992) and how do carnivores deal behaviourally with variation in plant information (e.g. Geervliet et al 1998a)? How do plant–carnivore interactions affect the fitness and dynamics of all tritrophic participants?

Until recently both these approaches have been limited to single species plant–carnivore interactions, ignoring food web complexity (e.g. non-linear trophic interactions such as intra-guild-predation) and neglecting the impact plant–carnivore interactions can have on other species in the community. Recently, however, there has been increased attention on a more holistic community level approach to plant–carnivore interactions. Chemically mediated interactions between all members of plant-inhabiting arthropods are being investigated as a basis for studying the importance of indirect food web interactions at the community level (e.g. Pallini et al 1997). It is realized that inferences on the evolution of this indirect mode of defence are tenuous when the emission of these volatiles is not viewed in a broader ecological context of species interactions. After all, the emitted herbivore-induced plant volatiles are 'free information' that can also be exploited by organisms other than the beneficial carnivores. Responses of herbivores, other plants, competitors or predators of the carnivores, may also determine the ecological costs and benefits of these plant signals (for a discussion on this see Dicke & Vet 1999).

Evolution of plant–carnivore interactions

Although evolutionary models may predict the conditions under which plants would benefit from emitting signals to carnivores, and whether this trait is evolutionarily stable (e.g. Sabelis & De Jong 1988), this tells us little about the evolutionary history of plant–carnivore interactions. It is most likely that the induction and release of plant volatiles after herbivory have been tailored to serve other functions than to attract 'bodyguards'. Turlings et al (1995), for example, suggest that the production and emission of terpenoids, produced by the plant in response to herbivore damage, may have evolved because terpenoids are deleterious to the feeding herbivores, hence serving the initial function of direct, not indirect, defence.

When volatile cues, produced by feeding damage, reliably accompany the presence of suitable herbivores, possibilities arise for carnivores to exploit this information. The well developed ability of associative learning, as shown for many insect parasitoids, has probably played an important role in the evolution of the initial responses of these carnivores to plant signals. Once the signal response has developed to the net benefit of both plant and carnivore, the stage is set for co-evolution. When plant fitness is linked to plant selection and foraging behaviour of the carnivores, plants will be selected to produce clear, detectable and reliable signals. In their turn, carnivores are selected to optimize their response to the plant signals.

Hence the fact that plants conspire with carnivores can have considerable evolutionary and ecological consequences for both partners, but also for the herbivore that is caught between the devil and the deep blue sea. Below I will consider some of these consequences in more detail.

The plant

Selection for good signalling ability can potentially affect all components of a plant's defence strategy: the composition of the odour blend, signal transduction and biosynthetic pathways, systemic responses to herbivore damage, or the relative investments into direct and indirect defence. Selection will act on existing genetic and phenotypic variation in these traits within and between species. The possibilities for change are obviously limited by certain phylogenetic and physiological constraints.

The existence of variation in the composition of odour blends is increasingly being shown for different plant systems. Plant volatiles may differ due to different abiotic factors, but also among plant species, plant genotypes, plant parts and according to different biotic factors such as infestation by different herbivore species, stages and densities (see Dicke 1998, Dicke & Vet 1999, Dicke 1999, this volume). Hence, there are, and were, possibilities for selection to alter the qualitative and quantitative composition of the plant odours produced after damage to meet the receiver's need for clarity. We see a great variety of plant responses to mechanical and herbivore damage (Fig. 1). Some plants (on the left of the figure) produce the same odour blend after mechanical or herbivore damage, i.e. there are no qualitative differences in odour composition (e.g. cabbage plants, Mattiacci et al 1994, Geervliet et al 1997). Plants at the other extreme (on the right) show a different strategy since here novel information dominates the volatile blend. The major components of the odour blend produced after damage by herbivores consists of compounds that are not emitted when plants are mechanically damaged (or undamaged) and different herbivore species induce specific volatile compounds (e.g. broad bean, Powell et al 1998).

Seen in the light of co-evolution between plants and carnivores there will be selection on the qualitative and quantitative composition of the odour blend to optimize signal clarity. Hence, the way carnivores perceive and process qualitative and quantitative differences in odour blends may direct the evolution of the plant signal in response to herbivore damage (see below).

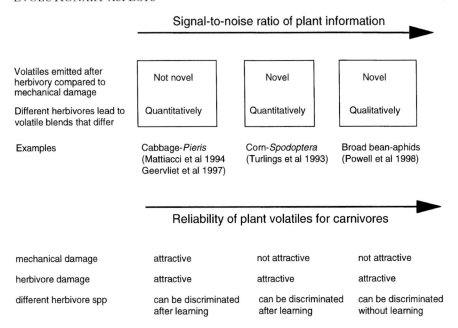

FIG. 1.   The relationship between signal-to-noise ratio of chemical plant information that is produced after damage and the response of carnivores.

The carnivore

In the evolutionary setting of real plant–carnivore mutualism, carnivores, in turn, are selected to optimize their response to those plant signals that indicate the presence of suitable herbivores. They can do this by increasing their sensory and behavioural ability to perceive the signal, by efficient information processing and decision making, by enhanced discrimination between signal and noise, and by other behavioural adaptations. I expect carnivores to show some degree of adaptation to the most prevailing and suitable plant–herbivore complexes. Hence, carnivore populations are expected to vary in their responses to plant volatiles and other major foraging cues. Kester & Barbosa (1991) compared different populations of the parasitoid *Cotesia congregata* that were associated with *Manduca sexta* on tomato and tobacco. The populations not only differed in survivorship responses to nicotine concentrations in their host's diet, but also in ovipositional preferences for hosts on tobacco or tomato in the field and in searching responses to these host plants. Unfortunately, there is still very little knowledge of such population differences and of genetic variation in the use of plant volatiles, or infochemicals in general, by predators and parasitoids. In contrast, at least for parasitoids, there is more insight (and more studies) into phenotypic plasticity in the use of infochemicals. Perhaps this is not surprising with respect to plant cues. The variable nature of plant cues may hamper the evolution of genetically fixed responses, especially in longer-living carnivores. Selection for flexibility in response to plant cues can be expected. Indeed we do find much phenotypic plasticity, due to learning processes, in the responses of parasitoids to plants (Turlings et al 1993, Vet et al 1995).

Returning to Fig, 1, selection will act on the carnivore's ability to discriminate between signal and noise, i.e. between relevant and irrelevant variation in plant information. Plant species differ in the specificity of information they provide to carnivores after being damaged in different ways: mechanical damage, damage by non-suitable herbivores and damage by suitable herbivores. The more specific the information on the presence and identity of suitable herbivores, the easier it will be for carnivores to fine-tune their sensory and behavioural responses to these cues, either by genetic adaptation or by phenotypic plasticity.

Signal-to-noise ratio of plant information and responses of carnivores

We can hypothesise that qualitative differences are more important in odour recognition and discrimination learning than quantitative differences. Support for this hypothesis comes from experiments by Vet et al (1998). *Leptopilina heterotoma*, a parasitoid of *Drosophila* larvae that feed on different substrates, learned qualitative differences in odour cues more easily (i.e. with a simpler learning paradigm) than quantitative differences. On the left of Fig. 1 the signal-to-noise ratio of the plant information is low. No qualitatively different compounds are being emitted when plants are damaged mechanically or by different herbivore species. Hence, any type of damage in the field will create a considerable amount of noise, i.e. plant volatiles not related to the presence of suitable herbivores. An example of such a plant is cabbage (Mattiacci et al 1994, Geervliet et al 1997). Both mechanically damaged plants and plants damaged by non-host herbivore species are attractive to the parasitoid *Cotesia glomerata* (Steinberg et al 1993, Geervliet et al 1998b) and the lack of specificity in chemical plant information is a likely constraint on the optimization of foraging by this parasitoid species.

The signal-to-noise ratio increases when shifting to the right in Fig. 1. Plants emit novel compounds after being damaged by herbivores and (total right) even produce different compounds after being damaged by different herbivores.

We expect that the signal-to-noise ratio of the information influences the way carnivores respond. When signal-to-noise ratio is low, learning of small but consistent quantitative differences in odour blends or the use of other sensory modalities will be required to discriminate between signal and noise. Empirical data for *C. glomerata* show that during foraging this parasitoid does acquire the ability to distinguish even between plants infested by one of two *Pieris* host species that differ in suitability (Geervliet et al 1998b). It remains unknown whether this acquired ability is based on the use of minor quantitative differences in odour blends or on the use of visual cues. *C. glomerata* may use visual cues, such as the differences in size of the feeding holes, to distinguish between host plants infested with *P. brassicae* or *P. rapae*. In behavioural experiments using cabbage plants with artificial holes of different sizes, the parasitoids were shown to readily associate the size of the holes with the presence of hosts (J. M. Verdegaal, F. L. Wäckers & L. E. M. Vet, unpublished data).

When signal-to-noise ratio is higher (Fig. 1, middle), discrimination between plants damaged mechanically or by herbivores is more readily achieved. The parasitoid *Cotesia marginiventris* is not attracted to mechanically damaged maize. Parasitoids are attracted to host-infested maize plants on the basis of herbivore-induced terpenoids that are not produced after mechanical damage. Plants infested by different herbivores do not differ in these induced terpenoids, but consistent quantitative differences exist in green leaf volatiles emitted as a result of mechanical damage. These differences appear to be sufficient for the parasitoids to associatively learn to distinguish between plants infested by one of two host species (Turlings et al 1993). On the right in Fig. 1 the signal-to-noise ratio is highest and learning may not be necessary for discrimination between plants infested with suitable and non-suitable herbivores. Empirical data show that the aphid parasitoid (*Aphidius ervi*) readily discriminates between broad bean plants infested with a suitable and a non-suitable aphid host and that experience is not needed for this discrimination. Plants infested with the non-host aphid are not even preferred over undamaged plants (Du et al 1996, Powell et al 1998). Similar evidence is given by Takabayashi et al (1995) for discrimination among plants infested by herbivores of different larval instars. Without prior oviposition experience, the parasitoid *Cotesia kariyai* is attracted to maize plants infested by 1st–4th instar caterpillars (suitable host instars), while plants infested by 5th–6th instar larvae (non-suitable host instars) are not attractive. The same results are obtained when plants are treated with regurgitant of 3rd instar larvae (attraction) or 6th instar larvae (no attraction). I refer to Dicke & Vet (1999) for further examples.

Is specificity of plant information synonymous with qualitative differences?

Although many studies have shown that parasitoids learn plant odours (reviewed in Turlings et al 1993, Vet et al 1995) it remains to be elucidated *how* parasitoids deal behaviourally with natural variation in plant odour cues. We know that *Cotesia glomerata* learns to discriminate between odours of different cabbage cultivars (Geervliet et al 1998b). The question remains *what* these parasitoids learn and on what basis they discriminate between these cabbage plants. Do they learn complete odour bouquets, or do they learn on the basis of qualitative differences in one (or a few) key components between plants? The latter is perhaps suggested by studies with *L. heterotoma* where parasitoids easily learned to distinguish between yeast patches that differed in the presence of one volatile compound only (Vet et al 1998). Note that we go beyond the more traditional question of how insects (mostly herbivores) recognize and select plants on the basis of chemical information (Bernays & Chapman 1994). Here we focus on how insects deal sensorily and behaviourally with *variation* in plant volatiles.

If it is indeed easier for carnivores to perceive, process and learn qualitative differences, and if qualitative differences are interpreted as synonymous with specificity, this may direct the evolution of the plant signal in response to herbivore damage towards greater qualitative differentiation of the odour blend.

The herbivore

The very fact that plants guide the searching of carnivores has evolutionary and ecological consequences for the herbivore. The conspiracy between plant and carnivore is based on the feeding activity of the herbivore itself and selection will act on the herbivore's diet choice and feeding behaviour to reduce the conveyance of information that reveals its presence. The elicitor of plant volatiles has been identified as a component of herbivore oral secretions (Mattiacci et al 1995, Alborn et al 1997). Selection may act on variation in the quality and quantity of oral secretions to minimize the induction of volatile production. Another strategy to sabotage the effect of induced defence is to desert locations emitting volatiles (such as feeding holes), to continue feeding on a different part of the leaf (Mauricio & Bowers 1990), or to chew the leaf petiole before consuming the detached leaf (Weinstein 1990).

Herbivores can select food plants that constitute a relatively enemy-free space. Escape from carnivores can be a driving force in the evolution of plant selection by herbivores (e.g. Bernays & Graham 1988). Plant selection may involve e.g. the choice of plant species, plant genotypes, plants under different abiotic conditions, and plant parts. The butterfly *Pieris napi japonica*, for example, prefers to oviposit on *Arabis spp.* while these plants are of lower quality for larval development than alternative food plants. However, it was demonstrated that the *Arabis* plants were representing an enemy-free space since they were searched less readily by the herbivore's major parasitoid, *C. glomerata* (Ohsaki & Sato 1994).

The chances of a herbivore successfully escaping from carnivores through plant selection are best when there is a relative focus of the carnivores on plant stimuli over herbivore stimuli, i.e. for carnivores specialized at the plant level (Vet & Dicke 1992). A successful shift will then depend on the chemical relatedness of the induced odours of the two plant species and the way carnivores use the odours. Chemical relatedness of plant volatile profiles is determined by the specifics of sensory perception. Insects such as bees are known to generalize between different classes of odours which apparently are interpreted as chemically related (e.g. Getz & Smith 1990). However, as mentioned above, little is known about how carnivorous arthropods perceive and interpret plant odour composition and its variation.

Herbivore density affects the amount of induced volatiles and thus the degree of carnivore attraction. If plants do not respond to a low herbivore load or if carnivores ignore the response of plants to a low herbivore density, such a density also reflects an enemy-free space. Hence, even the evolution of clutch size decisions of herbivores may be influenced by plant–carnivore interactions.

Epilogue

The conspiracy between plants and carnivores is in the heart of evolutionary ecology, and is wide open for rigorous experimental and theoretical investigations. It is, for example, important to study whether the laboratory-measured mechanisms of induced defence really act as protecting systems under realistic field conditions. There is promising progress towards the manipulation of plant traits by means of the plant's signal transduction pathway (Karban & Baldwin 1997), providing exciting experimental possibilities to study the costs of plant volatile production, measured in terms of plant fitness. The elegant field experiments of Baldwin (1998) have shown the power of using this experimental approach to study the ecology of direct defence.

As mentioned above, the way carnivores deal with variation in plant odours in nature remains a striking gap in our knowledge. Experimental behavioural analyses with natural and modified odour blends can help us assess whether quantitative or qualitative differences in compounds matter to foraging carnivores.

To finish, it goes without saying that studies on defence in plants are of great value to the applied field of insect pest management. After all, plant defence is central to plant resistance (direct defence) and biological control (indirect defence), two major pillars of the promising practice of integrated pest management.

References

Alborn T, Turlings TCJ, Jones TH, Steinhagen G, Loughrin JH, Tumlinson JH 1997 An elicitor of plant volatiles from beet armyworm oral secretion. Science 276:945–949
Baldwin IT 1998 Jasmonate-induced responses are costly but benefit plants under attack in native populations. Proc Natl Acad Sci USA 95:8113–8118

Bernays EA, Chapman RF 1994 Host–plant selection by phytophagous insects. Chapman & Hall, New York

Bernays EA, Graham M 1988 On the evolution of host specificity in phytophagous arthropods. Ecology 69:886–892

Dicke M 1994 Local and systemic production of volatile herbivore-induced terpenoids: their role in plant–carnivore mutualism. J Plant Physiol 143:465–472

Dicke M 1998 Evolution of induced indirect defence of plants. In: Tollrian R, Harvell CD (eds) The ecology and evolution of inducible defences. Princeton University Press, Princeton, NY, p 62–88

Dicke M 1999 Specificity of herbivore-induced plant defences. In: Insect–plant interactions and induced plant defence. Wiley, Chichester (Novartis Found Symp 223) p 43–59

Dicke M, Vet LEM 1999 Plant–carnivore interactions: evolutionary and ecological consequences for plant, herbivore and carnivore. In: Olff H, Brown VK, Drent RH (eds) Herbivores: between plants and predators. Blackwell Science, Oxford, p 483–520

Du Y-J, Poppy GM, Powell W 1996 Relative importance of semiochemicals from first and second trophic levels in host foraging behavior of *Aphidius ervi*. J Chem Ecol 22:1591–1605

Geervliet JBF, Vet LEM, Dicke M 1996 Innate responses of the parasitoids *Cotesia glomerata* and *C. rubecula* (Hymenoptera: Braconidae) to volatiles from different plant–herbivore complexes. J Insect Beh 9:525–538

Geervliet JBF, Posthumus MA, Vet LEM, Dicke M 1997 Comparative analysis of headspace from different caterpillar-infested or uninfested food plants of *Pieris* species. J Chem Ecol 23:2935–2954

Geervliet JBF, Ariëns SJ, Dicke M, Vet LEM 1998a Long-distance assessment of patch profitability through volatile infochemicals by the parasitoids *Cotesia glomerata* and *Cotesia rubecula* (Hymenoptera: Braconidae). Biol Control 11:113–121

Geervliet JBF, Vreugdenhil AI, Dicke M, Vet LEM 1998b Learning to discriminate between infochemicals from different plant–host complexes by the parasitoids *Cotesia glomerata* and *Cotesia rubecula* (Hymenoptera: Braconidae). Entomol Exp Appl 86:241–252

Getz WM, Smith KB 1990 Odorant moiety and odor mixture perception in free-flying honey bees (*Apis mellifera*). Chem Senses 15:111–128

Godfray HCJ 1995 Communication between the first and third trophic levels: an analysis using biological signalling theory. Oikos 72:367–374

Hopke J, Donath J, Blechert S, Boland W 1994 Herbivore-induced volatiles: the emission of acyclic homoterpenes from leaves of *Phaseolus lunatus* and *Zea mays* can be triggered by a β-glucosidase and jasmonic acid. FEBS Lett 352:146–150

Hunter MD, Price PW 1992 Playing chutes and ladders: heterogeneity and the relative roles of bottom-up and top-down forces in natural communities. Ecology 73:724–732

Karban R, Baldwin IT 1997 Induced responses to herbivory. University of Chicago Press, Chicago, IL

Kester KM, Barbosa P 1991 Behavioral and ecological constraints imposed by plants on insect parasitoids: implications for biological control. Biol Control 1:94–106

Mattiacci L, Dicke M, Posthumus MA 1994 Induction of parasitoid attracting synomone in Brussels sprouts plants by feeding of *Pieris brassicae* larvae: role of mechanical damage and herbivore elicitor. J Chem Ecol 20:2229–2247

Mattiacci L, Dicke M, Posthumus MA 1995 β-glucosidase: an elicitor of herbivore-induced plant odor that attracts host-searching parasitic wasps. Proc Natl Acad Sci USA 92:2036–2040

Mauricio R, Bowers MD 1990 Do caterpillars disperse their damage? Larval foraging behaviour of two specialist herbivores, *Euphydras phaeton* (Nymphalidae) and *Pieris rapae* (Pieridae). Ecol Entomol 15:153–161

Maynard Smith J 1982 Evolution and the theory of games. Cambridge University Press, Cambridge

Ohsaki N, Sato Y 1994 Food plant choice of *Pieris* butterflies as a trade-off between parasitoid avoidance and quality of plants. Ecology 75:59–68

Pallini A, Janssen A, Sabelis MW 1997 Odor-mediated responses of phytophagous mites to conspecific and heterospecific competitors. Oecologia 100:179–185

Powell W, Pennacchio F, Poppy GM, Tremblay E 1998 Strategies involved in the location of hosts by the parasitoid *Aphidius ervi* Haliday (Hymenoptera: Braconidae, Aphidiinae). Biol Control 11:104–112

Price PW, Bouton CE, Gross P, McPheron BA, Thompson JN, Weis AE 1980 Interactions among three trophic levels: influence of plant on interactions between insect herbivores and natural enemies. Annu Rev Ecol Syst 11:41–65

Sabelis MW, De Jong MCM 1988 Should all plants recruit bodyguards? Conditions for a polymorphic ESS of synomone productionin plants. Oikos 53:247–252

Steinberg S, Dicke M, Vet LEM 1993 Relative importance of infochemicals from first and second trophic level in long-range host location by the larval parasitoid *Cotesia glomerata*. J Chem Ecol 19:47–59

Strong DR, Lawton JH, Southwood R 1984 Insects on plants: community patterns and mechanisms. Harvard University Press, Cambridge, MA

Takabayashi J, Takahashi S, Dicke M, Posthumus MA 1995 Developmental stage of herbivore *Pseudaletia separata* affects production of herbivore-induced synomone by corn plants. J Chem Ecol 21:273–287

Tumlinson JH, Turlings TCJ, Lewis WJ 1993 Semiochemically mediated foraging behavior in beneficial parasitic insects. Arch Insect Biochem Physiol 22:385–391

Turlings TCJ, Tumlinson JH, Lewis WJ 1990 Exploitation of herbivore-induced plant odors by host-seeking parasitic wasps. Science 250:1251–1253

Turlings TCJ, Wäckers FL, Vet LEM, Lewis WJ, Tumlinson JH 1993 Learning of host-finding cues by hymenopterous parasitoids. In: Papaj DR, Lewis AC (eds) Insect learning: ecological and evolutionary perspectives. Chapman & Hall, New York, p 51–78

Turlings TCJ, Loughrin JH, McCall PJ, Röse USR, Lewis WJ, Tumlinson JH 1995 How caterpillar-damaged plants protect themselves by attracting parasitic wasps. Proc Natl Acad Sci USA 92:4169–4174

Vet LEM, Dicke M 1992 Ecology of infochemical use by natural enemies in a tritrophic context. Annu Rev Entomol 37:141–172

Vet LEM, Wäckers FL, Dicke M 1991 How to hunt for hiding hosts: the reliability-detectability problem in foraging parasitoids. Neth J Zool 41:202–213

Vet LEM, Lewis WJ, Cardé RT 1995 Parasitoid foraging and learning. In: Cardé RT, Bell WJ (eds) Chemical ecology of insects 2. Chapman & Hall, New York, p 65–101

Vet LEM, De Jong AG, Franchi E, Papaj DR 1998 The effect of complete versus incomplete information on odor discrimination in a parasitic wasp. Anim Behav 55:1271–1279

Weinstein P 1990 Leaf petiole chewing and the sabotage of induced defences. Oikos 58:231–233

DISCUSSION

Poppy: I'm interested in the question of qualitative versus quantitative change. Louise Vet, like myself, you view a qualitative change as a new compound coming into the picture. But I think some of the people in the audience here can view qualitative change as a dramatic change in the ratio of existing compounds. What is the general consensus about this? And if one is talking about changes in the ratio, how dramatic do these changes need to be to separate them from the inherent variability in chemical profiles from undamaged and damaged plants?

Pickett: As a chemist, I have always regarded qualitative differences as involving the generation of new compounds. But, as you say, as far as the insect or any other recipient of semiochemical messages is concerned, a change in ratio could give a quality signal.

Firn: How many compounds are involved in detection? When you are talking about a ratio, are you talking about just two compounds? How many bits of information from individual compounds does an insect need to be able to make a choice?

Vet: That is a critical question. When an insect learns to respond to a novel plant, we have no idea precisely what it actually learns. For instance, does it learn the whole composition of plant odours, consisting of many compounds in different ratios? I would like to hear what plant physiologists or insect physiologists have to say about this.

Firn: Do insects respond to a loss of information as well as a gain? If you remove a volatile, does this put the insect off?

Vet: Yes. In bioassays if you try to make a mixture of compounds that will give a similar response as the actual plant, then although there may be a response it is frequently not as strong as to the genuine volatile composition.

Tumlinson: The number of compounds involved in this sort of interaction depends on the plant: in some you get fewer compounds and others, such as cotton, there is an enormous number involved. In the early work that Ted Turlings did with me in which we were trying to duplicate the volatiles produced by corn (Turlings et al 1991), we could make a synthetic blend that attracted the parasitoids, but it didn't compete with the natural blend — they could tell the difference, even though all the same compounds were present, simply because it's very difficult to reproduce and dispense exactly the same proportions of compounds in a multi-component blend. I therefore think that small differences in proportions of compounds are going to be detected and learned by these insects. When we start talking about qualitative differences, these can be very subtle.

Turlings: The differences that we observe in the volatile blends by chromatography are not that important: instead, what is critical is whether or not the insects can smell the difference. If you look at two closely related species feeding on corn plants, there are hardly any differences in the chromatograms except for one compound that's produced a lot more by one than the other. But I can easily smell the difference and, after some experience, just like the parasitic wasp, tell you which species it is. The same thing is true with the synthetic blend: this looks almost identical to the natural blend, but it smelt quite different to me as well to the insect. But if you give the insect the appropriate experience, it is very much attracted to it.

Vet: You are putting an important qualifier in here: 'if you give the insect the appropriate experience'. It needs that experience, but then what does it learn? In the case where there are two odour blends that are practically identical, but only differ a little in one component, your conclusion must be that they find out about that. But do they find out about the one component only and the rest has to be exactly the same, or is it that the rest doesn't really have to be exactly the same — it could be variable — but it's not noticed as being an important variable?

Städler: There is a technical problem here, also. We always think that the GC chromatogram is representing what the insect is smelling. We have had quite a few examples where the chromatogram doesn't show what is really essential, because the critical components are present in such tiny amounts or because the column did not let these compounds go through.

Tumlinson: There's no doubt that the gas chromatograph is not nearly as sensitive as the insect at detecting compounds and differences.

Dicke: Our appreciation of what the plant does has steadily been growing each year. For instance, with the lima bean system with spider mites, 10 years ago we had some 17 compounds. Subsequently our equipment has improved and now we can detect four times as many compounds. We are getting more knowledgeable about the minor components of the volatile blend. Thus our appreciation of the complexity that the insect is confronted with has increased with technological advances.

Firn: Presumably in 10 years' time you will be seeing hundreds of volatiles in the GC traces as detection sensitivity increases. The question is, which of these components are the important ones? Has anyone done any work with mutants which lack particular volatiles, to see whether that actually modifies the insect responses?

Dicke: I am not aware of work with mutants, but research here in Britain at the John Innes Centre has made two selected lines of *Brassica oleracea* that differ in glucosinolate composition and thus in volatiles derived thereof. They have shown effects on the behaviour of the parasitoid *Diaretiella rapae* (Bradburne & Mithen 1998).

Tumlinson: Insects are learning, and with just a few exceptions I don't think that there's any one compound that's vital: it is the whole blend the response is addressed to. Therefore, if you remove one compound, the insects will learn the blend that's remaining.

Vet: Thinking in evolutionary terms, what then is a better signal? If there is selection pressure on the plants for reliable signals, if there can be selection on the qualitative/quantitative differences in the plant, is it going to be a qualitative difference, or is it just as good to be quantitatively different?

Haukioja: If the evolutionary role of the production of signals by the plant is to reduce herbivory, and the parasitoids need learning to be able to connect their behaviour to this signal, isn't it a surprisingly complicated mechanism? I mean that the first time the parasitoid chooses the herbivore it does it by random and then it learns, and only after learning will the signal really be of benefit to the plant.

Poppy: But there are demonstrations of innate responses. We have shown with the aphid parasitoid, *Aphidius ervi*, that there are innate responses to some of these herbivore-induced synomones (Du et al 1996), and Professor van Emden (Reading University, UK) has suggested that Hopkins' Host Selection Principle may apply to aphid parasitoids, although he now believes that their results may be explained by early adult learning, when the parasitoids emerge from the mummy (Wickremasinge & van Emden 1992, van Emden et al 1997). You could say therefore that there is likely to be no such thing as a truly naïve parasitoid, because wherever it emerged from it has gained chemical information from the host and the host plant.

Tumlinson: But as soon as the host moves to another plant, it has to learn again. This may benefit it immediately, but if there are no other plants of that same type or those same odours around, the learning process has to start over again.

Vet: I think the function of the learning process is to cause a real reduction in spatiotemporal variation. The parasitoid enters a situation that is highly unpredictable beforehand, so by being able to focus on that plant type and the surrounding plant population, the parasitoid will get a benefit in its foraging efficiency. Whether or not the plant benefits from that is not as clear: there are many factors involved. It could be, of course, that by having encountered one caterpillar where there is an aggregated distribution of caterpillars, it will benefit because the parasitoid will remain in that area. Or if plants are related to other plants that are there, closely related plants will benefit because they are more related then the plants that are further away: that's where you can see a benefit of the learning for the plant.

Karban: What is the evidence that plants benefit from releasing this information?

Turlings: I'll try to answer this question in my paper (Turlings & Fritzsche 1999, this volume), but it's not the ultimate evidence. We found that parasitization by *Cotesia marginiventris* reduces feeding by its hosts tremendously: far less leaf material was eaten by parisitized than by unparasitzed caterpillars. In a preliminary laboratory test, this resulted in a higher survival rate of plants attacked by parasitized caterpillars. For other parasitoids this may be quite different. As Louise Vet mentioned, there are even parasitoids that cause their hosts to eat more.

Poppy: The aphid situation is an unusual one, because we are dealing with a clonal organism and a species which will reproduce parthenogenetically and has telescopic generations. Therefore, from an evolutionary point of view the benefit of the plant calling in a parasitoid to kill an aphid are clear: it can have a big impact in terms of that population of aphids on that plant in a very short time. However, with Lepidoptera this may not always be the case: after all, what is the benefit to a plant of bringing in a parasitoid that increases the duration of feeding? That individual plant hasn't benefited at all in the short term.

Vet: I tried to show with the data from my postdoc, Jeffrey Harvey, that this is not really the case. He showed that there was a relation between feeding rate of parasitized caterpillars and parasitoid clutch size, but then it depends whether or not this is more than the feeding rate of an unparasitized caterpillar: this is the crucial fact.

Haukioja: If production of these volatiles is an evolved defence, there should be (or should have been) genetic variance in volatile production due to insect damage. Is there any information on that?

Dicke: Genotypes of plants differ in what they emit. This is as far as we got. No one has shown that you can select for differences, apart from perhaps the research at the John Innes Centre I just mentioned, and the evidence from Jim Tumlinson's group that naturalized cotton produces more volatiles than agricultural varieties. This might indicate that in the absence of selection on this production you lose it. However, this is indirect evidence.

Tumlinson: One problem is that too many of us are working with agricultural varieties. As Marcel Dicke said, we examined about five different varieties of cotton that are grown in southeastern USA. There is slight variation, but not a whole lot. But cotton that is naturalized in the Everglades showed a huge difference: it releases probably eight times the quantity of volatiles.

Vet: Did you follow-up that with field experiments showing the effects of these differences on insects?

Tumlinson: No.

Dicke: With all the work on agricultural plants, it is perhaps surprising to see that in virtually all agricultural plants we have studied we find this phenomenon. One might expect that volatile release would not have been selected for because these crops have always been bred for other traits, such as yield or disease resistance. This selection might even have worked against volatile production.

Pickett: I'd like to say something about non-agricultural crops. In the example I gave in the introduction of the stem-borer control in Africa (p 1), we actually use non-food crops for the intercropping repellent crop plant and for the attractant trap crop. We are thereby capitalizing on the two extremes. We didn't think about it in these terms initially: we simply looked at a whole range of plants and found the ones, mostly wild or forage plants, that did what we required of them.

Schultz: This *segues* to another question that Louise Vet raised about early and late succession. As I look around the room I think there are only a couple of us who work late in succession, on trees for instance. I would prefer not to argue from absence of evidence about whether these phenomena are important late in succession. There are plenty of natural history hints to suggest that they probably are operating. For example, many tree-feeding insects sniff the leaves and move on a basis that suggests that they're trying to avoid being the target. I think we over-simplify our view of the chemistry of late successional plants: they are more complex than we give them credit for. Secondly, there just hasn't been enough work done to draw a conclusion about this. I'll bet we're going to find all of the same phenomena.

Vet: Interestingly, we have a lot of generalist parasitoids there and we know from one study in which we compared the behaviour of generalists and specialists, that the generalists in particular seem to do a lot of learning, because they can fill in the details as they go along.

Schultz: Of course, with trees we have the handicap that if we're going to answer Rick Karban's question about whether there is an advantage to the plant, we have a 50-year wait for them to reproduce.

Pickett: When the plant is damaged, it produces a rapid burst of a large number of compounds. As Ted Turlings says, the insect may not be responding to the major components. This then takes us back to the question that Richard Firn raised: what is the plant making all that material for? How has it evolved to do that? Is the volatile release a side issue to other more important defence chemistry associated with the oxidative bursts, or are the associated chemicals having some effect outside the particular system with which we are concerned, for example against pathogens rather than insects?

Firn: While a plant may benefit directly by making a few compounds to modify insect behaviour, it would be wrong to imagine that we must seek a role for every volatile compound produced. There are good reasons to expect considerable redundancy (Firn & Jones 1996). Because there is a low probability of any chemical possessing specific, potent biological activity, plants will have been selected which produce a great diversity of volatiles. There will probably be many compounds that are along for the ride. This is a reasonable explanation of why there may be 100 volatile compounds produced when many fewer could give the quality of information necessary.

Schultz: I think that we're still waiting for an explanation as to why oak trees emit many millions of metric tons of isoprene in the atmosphere. Perhaps some of these things are combinatorial chemistry advantages that arise secondarily.

Lamb: Is there a correlation between chemical complexity and length of life cycle?

Schultz: There's not enough work done on the long life cycle end to answer that question.

Firn: Please remember that considerable chemical complexity occurs in microbes, which have an incredibly short life cycle. It is also worth noting that although we tend to think of volatile production as being unique to plants, it is common to many organisms. I think we give off volatiles if we are damaged, don't we?

Pickett: Yes. You can smell blood very sensitively. If I cut my hand now, you would probably smell it.

Firn: I had the impression that parasites were fairly specific. Therefore, is there a danger that parasitic insects are attracted to volatiles that have been induced by a herbivore that the parasite cannot use as a host?

Vet: You are right that parasitoids can be extremely specific. What happens is that the more specific these parasitoids are, the more directed or guided their search is by cues that are directly derived from the host itself.

Schultz: At the other extreme, my student, Ahnya Redman, has been able to show that Gipsy moth wounding of aspen trees attracts generalized parasitoids that then parasitize all the other caterpillar species in the same place. Apparently, volatiles emitted by the wounded trees attract a wide range of parasitoid species with broad host preferences.

Pickett: Jon Gershenzon, would you say that the enzymology backs up the idea that Richard Firn has suggested, that there is a more general production capability of volatiles and that is in fact what we are discussing? The enzymology does not seem highly specific. Many of the genes that have been cloned and then expressed for depyrophosphorylases can give a wide range of products. This would suggest the evolution of product generation is the issue, rather than of the production of specific compounds.

Gershenzon: By and large we don't have the information yet to know how many enzymes are involved in the production of herbivore-induced volatiles. First, we must remind ourselves that this is a large group of compounds which keeps getting larger with the increasing sensitivity of detection techniques. The production of some volatiles is induced by herbivory, whereas in certain plants, such as cotton (as shown by the work of Jim Tumlinson), there is actually a stored reservoir of preformed isoprenoid volatiles that are simply released by herbivory. These stored compounds, which are largely cyclic terpenoids, are actually sequestered in the sub-epidermal pigment glands and are released immediately by feeding. The herbivore-induced terpenoid compounds which emerge later are typically acyclic compounds. To produce these many different volatiles, a large number of different enzymes is probably required.

It is also interesting to note that almost every plant that has been looked at seems to produce a rather similar set of herbivore-induced volatiles. In the systems that have been studied, we frequently see linalool, ocimene, β-farnesene, caryophyllene, the C_{11} and C_{16} homoterpenes, etc. For some reason, we keep going back to the same set of compounds. This is in contrast to the compounds sequestered prior to attack, which show much more diversity. Thus, for the herbivore-induced volatiles, we may be dealing with the same enzyme systems in taxonomically distant species. This can help focus the search for the biochemical machinery involved. I think it is striking that, when you consider the enormous universe of isoprenoids out there—30 000 or so at last count—there is actually a rather restricted group that keep coming up again and again as herbivore-induced volatiles. In my mind, that makes the quest for the enzymology and function of these compounds much more interesting.

References

Bradburne R, Mithen R 1998 A single gene in *Brassica oleracea* that controls the attraction of the aphid parasitoid *Diaretiella rapae*. Abstracts 10th International Symposium on Insect–Plant Relationships, 4–10 July 1998, Oxford, UK, p 49

Du Y-J, Poppy GM, Powell W 1996 Relative importance of semiochemicals from the first and second trophic levels in host foraging behavior of *Aphidius ervi*. J Chem Ecol 22:1591–1605

Firn RD, Jones CG 1996 An explanation of secondary product redundancy. In: Romeo JT, Saunders JA, Barbosa P (eds) Phytochemical diversity and redundancy in ecological interactions. Plenum, New York (Recent Adv Phytochem 30) p 295–312

Turlings TCJ, Fritzche ME 1999 Attraction of parasitic wasps by caterpillar-damaged plants. In: Insect–plant interactions and induced plant defence. Wiley, Chichester (Novartis Foundation Symposium 223) p 21–38

Turlings TCJ, Tumlinson JH, Heath RR, Proveaux AT, Doolittle RE 1991 Isolation and identification of allelochemicals that attract the larval parasitoid, *Cotesia marginiventris* (Cresson), to the microhabitat of one of its hosts. J Chem Ecol 17:2235–2251

van Emden HF, Sponagel B, Baker T, Ganguly S, Douloumpaka S 1997 Hopkins' host selection principle: another nail in its coffin. Physiol Entomol 21:325–328

Wickremasinge MGV, van Emden HF 1992 Reactions of adult female parasitoids, particularly *Aphidius rhopalosiphi*, to volatile cues from the host plants of their aphid prey. Physiol Entomol 17: 297–304

Attraction of parasitic wasps by caterpillar-damaged plants

Ted C. J. Turlings and Maria Elena Fritzsche

Institute of Zoology, University of Neuchâtel, Rue Emile-Argand 11, CH-2007, Neuchâtel, Switzerland

Abstract. Plant volatiles emitted in response to herbivory have been suggested to function as signals to attract natural enemies of herbivores. Most known examples of induced plant volatiles used by natural enemies involve parasitoids that locate caterpillars by means of odours emitted by plants after caterpillar attack. We study the tritrophic system that comprises the parasitoid *Cotesia marginiventris*, host caterpillars from the genus *Spodoptera*, and maize plants. Among the volatiles emitted by caterpillar-damaged maize plants, sesquiterpenes and indole are particularly attractive to the parasitoid. The usefulness of these plant volatiles for parasitoids is obvious. Less clear is their benefit to plants that emit them, as in most cases parasitization does not immediately stop caterpillars from damaging plants. However, plants appear to benefit directly from attracting *C. marginiventris*, as parasitized caterpillars consume considerably less plant tissue than unparasitized caterpillars. It is expected that in systems where parasitoids significantly reduce herbivory, they have contributed to selective pressures that have shaped the phenomenon of herbivore-induced volatile emissions by plants.

1999 Insect–plant interactions and induced plant defence. Wiley, Chichester (Novartis Foundation Symposium 223) p 21–38

Many plants respond to herbivory with the production of defence chemicals (Karban & Baldwin 1997). Several of these induced substances are volatiles and result in odours that are readily used by natural enemies (predators and parasitoids) to locate herbivores. It has been suggested that one of the functions of induced volatiles is to serve as signals to attract natural enemies, as plants' fitness may benefit from the interaction (e.g. Dicke et al 1990a, Dicke 1995, Sabelis & De Jong 1988, Turlings et al 1995). However, the benefit to an individual plant of attracting parasitoids is not always clear, because parasitized caterpillars will in most cases continue to feed and cause significant damage to a plant. Here we will discuss several aspects of parasitoid attraction by means of herbivore-induced plant odours and use results from our own studies to argue that parasitoids contribute to selective forces that have resulted in and maintained the phenomenon of herbivore-induced plant volatile emission.

21

Caterpillar-induced plant volatiles

Changes in volatile emissions in response to caterpillar feeding are easily demonstrated with young maize plants. The resulting odour can even be detected with the human nose several hours after initial insect feeding. The volatiles that are involved can be collected on adsorbents and gas chromatographic analyses of extracts of these collected volatiles show dramatic differences between caterpillar-damaged and healthy plants (e.g. Turlings et al 1990, Takabayashi et al 1995). The response of the plant is triggered by elicitors in the oral secretion of caterpillars (Turlings et al 1993a). One such elicitor (volicitin) was identified from the regurgitant of *Spodoptera exigua* (Alborn et al 1997, Tumlinson 1999, this volume).

Figure 1 illustrates the temporal changes in volatile releases that are seen after a maize plant is mechanically damaged and the damaged sites are treated with caterpillar regurgitant (in this case from *S. littoralis*). At 9 am 8-day-old maize plants were treated and immediately after treatment volatiles were trapped on filters every 0.5 h for a period of 12 h (Turlings et al 1998). Analyses of the collected odours (Fig. 1) show that immediately after damage the plants emit typical green-leafy aldehydes, alcohols and acetates that 'bleed' directly from the damaged sites. Without any further damage the emission of these highly volatile compounds will cease within hours. After 2–3 h the first induced volatiles (monoterpenes and indole) are emitted. It takes a few hours more before the emission of sequiterpenes is observed (Fig. 1). The emission of the induced compounds lasts throughout the entire collection period, but slowly wanes after reaching a maximum late during the day. In the context of their role in defence against herbivores, the observed timing of emissions is expected to be fast enough to allow the plant to attract the natural enemies long before the herbivores have done significant damage to the plant.

Attraction of parasitoids by means of herbivore-induced plant volatiles

Table 1 lists the studies that have demonstrated the importance of herbivore-induced plant odours in host location by parasitoids. Relatively few systems have

---

FIG. 1. (*Opposite*)   Chromatographic profiles of odours collected from maize seedlings (var. Ioana) at different times after damage and treatment with caterpillar regurgitant. The numbers with the peaks represent: 1, (Z)-3-hexenal; 2, (E)-2-hexenal; 3, (Z)-3-hexen-1-ol; 4, (E)-2-hexen-1-ol; 5, (Z)-3-hexen-1-yl acetate; 6, linalool; 7, (3E)-4,8-dimethyl-1,3,7-nonatriene; 8, indole; 9, (E)-α-bergamotene; 10, (E)-β-farnesene; 11, (E)-nerolidol; 12, (3E, 7E)-4,8,12-trimethyl-1,3,7,11-tridecatetraene. Two remaining volatiles did not originate from the plants; pentadecane (pd) was released from the regurgitant that was used to treat the plants and imp. represents an impurity in the air. IS1 and IS2 are the internal standards *n*-octane and *n*-nonyl-acetate. Taken from Turlings et al (1998) with permission, copyright Springer-Verlag.

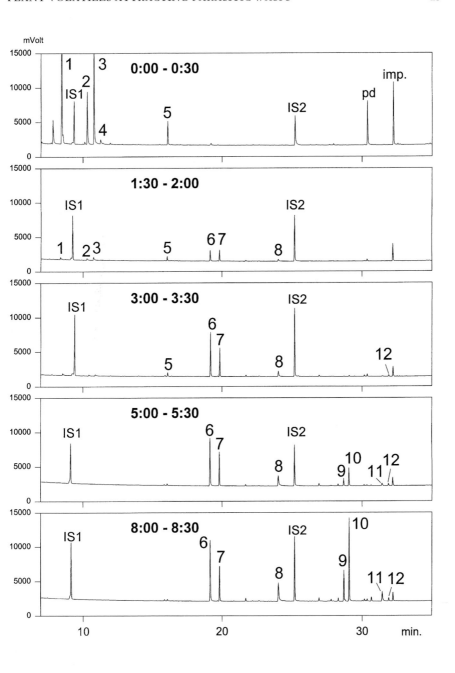

TABLE 1 Examples of parasitoids using caterpillar-induced plant odours to locate their hosts

Parasitoid	Herbivore(s)	Plant(s)	Specifics on the interactions	Selected references
Cotesia marginiventris	Spodoptera spp. Heliocoverpa zea Trichoplusia ni	Maize Cotton Cowpea	Plants respond to herbivory with the emission of volatiles that are attractive to the parasitoid. Factors in the regurgitant of the caterpillars elicit the emissions. One such elicitor has been identified and was named volicitin.	Turlings et al (1990, 1991a,b) Alborn et al (1997) Tumlinson (1999, this volume)
Microplitis croceipes	Heliocoverpa and Heliothis spp.	Cotton Cowpea Maize	The wasp is strongly attracted to volatiles emitted by the plants after they have been damaged by caterpillars or after treatment with caterpillar regurgitant.	McCall et al (1993) Turlings et al (1993a) Röse et al (1996)
Cotesia glomerata, Cotesia rubecula	Pieris spp.	Cabbage and related sub-species	The parasitoids are attracted to caterpillar-damaged plants due to an increase in odour emitted by the plant. β-glucosidase in the caterpillar oral secretion elicits this reaction in the plant.	Steinberg et al (1993) Agelopoulos & Keller (1994) Mattiacci et al (1994, 1995)
Cotesia kariyai	Pseudaletia separata	Maize	Maize plants eaten by early instar larvae of the herbivore show a significant increase in odour emissions and become attractive to the parasitoid. The plant does not respond to late instar larvae.	Takabayashi et al (1995)
Cotesia flavipes	Chilo partellus	Maize	The stem borer causes maize plants to systemically emit volatiles that attract the parasitoid.	Potting et al (1995)
Cardiochiles nigriceps	Heliothis virescens	Cotton Tobacco	Caterpillar feeding induces a specific emission of volatiles, which the parasitoid can distinguish from emissions induced by a non-host caterpillar.	De Moraes et al (1998)

been studied, but all show the importance of plant-provided signals, suggesting that the phenomenon is very common. These interactions have been mainly studied for parasitoids that attack Lepidoptera, but natural enemies that attack herbivores from other orders also make effective use of plant odours. One well studied example involves predatory mites that use induced plant odours to locate herbivorous mites, as discussed by Dicke (1999, this volume). In the case of mites, some of the individual compounds that are emitted by infested plants have been shown to be attractive to the predators (Dicke et al 1990b). For parasitoids, however, little is known about which of the plant volatiles are actually attractive.

Which of the compounds released from the plant are the most important in attracting parasitoids?

A commonly asked question concerns the importance of the individual plant-released volatile compounds for the attraction of parasitoids. The keen ability of parasitoids to learn any odour that is associated with the presence of host odour (Lewis & Tumlinson 1988, Turlings et al 1993b, Vet et al 1995) suggest that the entire blend released by a plant is important. Still, it can be expected that some compounds are more attractive (or better perceived) than others. This is the case for several specialist parasitoids. *Heydenia unica* is attracted to terpenoids associated with its host, the southern pine beetle (Camors & Paine 1972). Elzen et al (1984) found that several terpenoids released from cotton plants are particularly attractive to the parasitoid *Campoletis sonorensis* and Udayagiri & Jones (1992) showed that fractions of maize volatiles containing sesquiterpenes, aldehydes, a ketone and esters were much more attractive to the specialist parasitoid *Macrocentrus grandii* than polar fractions containing alcohols. The following experiment involving the parasitoid *Cotesia marginiventris* and its responses to volatiles emitted by caterpillar-damaged maize plants illustrates that it can be difficult to determine the relative importance of particular compounds.

   C. marginiventris is a solitary endoparasitoid that can develop in caterpillars of many different species. As a consequence, it can find suitable host caterpillars on many different plant species. One such plant is maize and a variety of experiments have shown that caterpillar-induced maize volatiles are very attractive to *C. marginiventris*. Maize volatiles are much more attractive than odours from the caterpillars or their faeces (Turlings et al 1991a). In one experiment the odours from maize seedlings damaged by caterpillars (*Spodoptera exigua*) were collected in large amounts and extracts of these collections were placed on strips of paper to test their attractiveness in a flight tunnel (Turlings et al 1991b). At an optimal dose, about 75% of female wasps tested were found to be attracted to the extract (Turlings et al 1991b). With this dose, an experiment was designed that allowed us

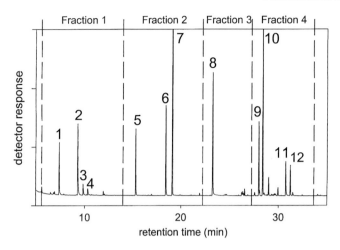

FIG. 2.   Chromatogram of volatiles collected in mass quantities from maize seedlings eaten by *Spodoptera* larvae. The four fractions that were taken for bioassays with the parasitoid *C. marginiventris* are indicated. Peak identities are given in the legend for Fig. 1.

to exclude parts of the total odour blend to determine how this affects the attractiveness of the extracts.

Odours of the complete plant–host complex were collected as described by Turlings et al (1991b). Using a technique first described by Brownlee & Silverstein (1968), we collected four fractions (see Fig. 2) of the total odour blend with a preparative gas chromatograph onto glass capillary tubes (see details in Turlings et al 1991b). In a first test, a recombination of all four fractions was compared with different recombinations of only three fractions. One at a time, recombinations were applied on a strip of paper (1×5 cm) as described by Turlings et al (1991b) and tested for attractiveness in a flight tunnel. Six 2-day-old experienced wasps (two groups of three) were allowed to fly to a single recombination and we recorded how many of them flew to the odour source. The recombinations were tested 'blind' in random order. The experiment was replicated five times on different days. Positive responses often resulted in landings on the paper strip, but a number of females approached a target strip within 3 cm without landing on it. Both these behaviours were recorded separately. The results in Fig. 3A show that the most attractive recombinations included fractions 1, 3 and 4. Surprisingly, this was more attractive than a recombination of all fractions.

Similar results were obtained with a choice experiment. The principal of this experiment was the same, except that all treatments were placed in the flight tunnel at the same time. This time only the four possible recombinations of three

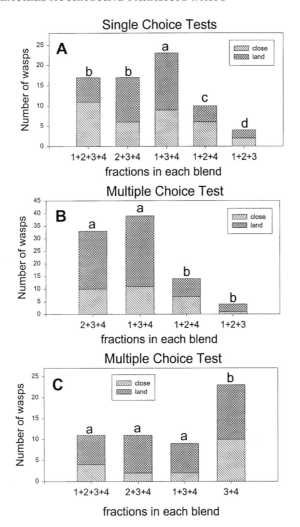

FIG. 3. Responses of wasps to recombinations of different odour fractions (see Fig. 2) placed on paper strips in a flight tunnel. (A) single choice tests, (B) and (C) multiple choice tests. Responses are divided between wasps that approached the target to within 3 cm (close) and wasps that landed on the target (land). The letters with each bar indicate significant differences in attractiveness among the test samples (Fisher's PLSD after ANOVA, $P < 0.05$).

fractions were used. They were applied on four different paper strips placed 10 cm apart perpendicular to the wind direction. The order in which they were placed was random. Groups of five experienced female wasps were released and their preferences (3 cm approaches and landings) were recorded. After each group of five the position of the paper strips was shifted one position to the right (the most right strip would move to the most left position), and a new group of five female wasps was released. This was repeated five times on a particular day and the entire experiment was repeated five times on different days. Again the recombination of fractions 1, 3 and 4 was the most attractive (Fig. 3B).

The main conclusion from this experiment is that fraction 2, which contains (Z)-3-hexen-1-yl acetate, linalool and (3E)-4,8-dimethyl-1,3,7-nonatriene, was the least attractive and may even have been repellent or masked the attractive fractions. This is surprising as these compounds do contribute significantly to the blend of induced volatiles. The results also indicated that fraction 1 did not contribute to the attractiveness of the entire blend.

One more experiment was conducted to further test the importance of fractions 1 and 2. For this, wasps were again offered four odours, one containing all four fractions, one without fraction 1, one without fraction 2, and one without both fractions 1 and 2. Most wasps flew to this last combination, which only included fractions 3 and 4 (Fig. 3C).

It is possible that the release rates of the highly volatile compounds in fractions 1 and 2 deviated too much from a natural release and thus cause repellence or masking of the attractive odours. It was previously found that when the entire blend was tested at very high concentrations its attractiveness also dropped significantly (Turlings et al 1991b). The negative effect of the two first fractions may therefore have been the result of limitations of the experimental procedure. Dose–response tests with these factions could perhaps show this. For now, it can be concluded that fractions that contain indole and several sesquiterpenes are highly attractive to the wasp.

Does a plant benefit from attracting parasitoids?

Parasitoid larvae do not kill their hosts immediately. The immature parasitoid develops within a living host, which in many cases continues to feed and grow. These so-called koinobiont parasitoids which allow further development of their host are common among larval parasitoids of Lepidoptera. Frequently, parasitization causes the caterpillars to eat even more than healthy caterpillars (e.g. Parker & Pinnell 1973, Byers et al 1993). In those cases, plants may not benefit directly from attracting parasitoids and it is unlikely that parasitoids contribute to selective pressures that favour plants to emit volatiles in response to herbivory. It seems, however, that there are numerous parasitoids that do reduce herbivory by their hosts and that they therefore may have direct positive effect on plant fitness (Beckage 1985). We are investigating this possibility for *C. marginiventris* and how this parasitoid may affect plant fitness.

One approach has been to compare the development of parasitized and unparasitized caterpillars (*Spodoptera littoralis*). Insects were kept under ambient

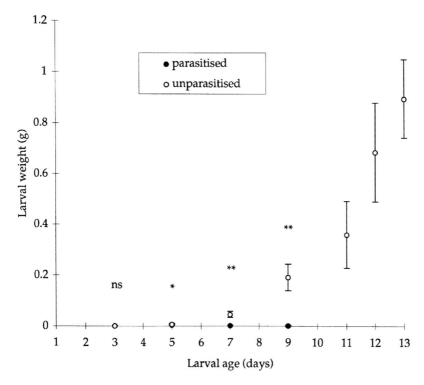

FIG. 4. Comparison between the weight gain of parasitized and unparasitized *Spodoptera littoralis* larvae. Six days after parasitization, larvae have gained very little weight. At this point the parasitoid larvae emerges from its host and the host dies, while unparasitized larvae continue to feed and grow considerably. The asterisks indicate significant differences in the weight of parasitized and unparasitized caterpillars (simple factorial ANOVA; ns, not significant; $*P < 0.02$; $**P < 0.0001$).

light (natural) and temperature in our laboratory. Three-day-old caterpillars that had been feeding on maize seedlings (var. Delprim) were offered to 2-day-old mated *C. marginiventris* females. Caterpillars that were observed to be stung by a parasitoid were removed and placed singly with an ample supply of maize leaves as food. Control (unparasitized) caterpillars that originated from the same egg batch were held the same way. Every two days the caterpillars were weighed and every day the leaves were replaced with fresh ones. The average weight gain of parasitized larvae, as shown in Fig. 4, was significantly reduced starting 2 days after parasitization. This agrees with previous reports on reduced consumption by caterpillars parasitized by *C. marginiventris* (Ashley 1983, Jalali et al 1988). The reduction in feeding results in less leaf damage (Table 2) and increases the chances

TABLE 2 Dry-weight of leaves from 5-week-old maize plants damaged by unparasitized or parasitized *Spodoptera litoralis* caterpillars

Treatment	Dry weight (g) of leaves (average ± SD)
Control plants ($n=16$)	0.78 ± 0.23
Plants eaten by unparasitized caterpillars ($n=16$)	0.43 ± 0.20
Plants eaten by parasitized caterpillars ($n=22$)	0.77 ± 0.17

Caterpillars (3 d old) were placed singly on the plants and the leaves were dried after pupa formation by the healthy caterpillars. Factorial analysis (ANOVA) showed a significant difference between treatments, $P < 0.0001$, $F = 17.081$.

of plants surviving a caterpillar attack (M. E. Fritzsche & T. C. J. Turlings, unpublished data).

Conclusions

Parasitoids make effective use of herbivore-induced plant volatiles to locate hosts for their offspring. The effects of secondary plant metabolites on insects that interact with plants have undoubtedly led to selection pressures that resulted in modifications in the production and release of these substances. The third trophic level of natural enemies of herbivores is likely also to have contributed to these selective pressures. Despite the fact that they do not immediately kill their host, many parasitoids significantly reduce plant damage caused by herbivores and thus may positively affect plant fitness. Plants that benefit from parasitoids are expected to have adapted the release of odours so as to optimize their attractiveness to these insects. However, this may be entirely different in cases where parasitoids and predators have no immediate effect on herbivory. If the attraction of the third trophic level is an important function of induced plant volatiles, comparative studies may reveal that induced plant signals occur principally in systems where plant fitness is positively affected by natural enemies of herbivores.

Acknowledgements

We thank Mark Scriber, Thomas Degen, and Betty Benrey for comments and suggestions. Much of what is discussed here was the result of research on *C. marginiventris* and its interactions with maize plants conducted by TCJT under the supervision of Dr Jim Tumlinson at the USDA-ARS, Gainesville, Florida and Dr Joe Lewis at the USDA-ARS, Tifton, Georgia. We are grateful for the weekly shipments of *Spodoptera littoralis* from Novartis, Pest Control, Basle, Switzerland. Dr Donald Nordlund and colleagues at the USDA-ARS, Knoxville, Mississippi, provided *Cotesia marginiventris*. Our current research is financed by The Swiss National Science

Foundation and a grant managed by The Swiss Centre for International Agriculture for the Swiss Agency for Development and Co-operation.

References

Agelopoulos NA, Keller MA 1994 Plant–natural enemy association in the tritrophic system *Cotesia rubecula–Pieris rapae–Brassicaceae* (Cruciferae). I: sources of infochemicals. J Chem Ecol 20:1725–1734

Alborn HT, Turlings TCJ, Jones TH, Stenhagen G, Loughrin JH, Tumlinson JH 1997 An elicitor of plant volatiles from beet armyworm oral secretion. Science 276:945–949

Ashley TR 1983 Growth pattern alterations in fall armyworm, *Spodoptera frugiperda*, larvae after parasitization by *Apanteles marginiventris*, *Campoletis grioti*, *Chelonomus insularis*, and *Eiphosoma vitticole*. Fla Entomol 66:260–266

Beckage NE 1985 Endocrine interactions between endoparasitic insects and their hosts. Annu Rev Entomol 30:371–413

Brownlee RG, Silverstein RM 1968 A micro-preparative gas chromatograph and a modified carbon skeleton determinator. Anal Chem 40:2077–2079

Byers JR, Yu DS, Jones JW 1993 Parasitism of the army cutworm, *Euxoa auxiliaris* (Grt.) (Lepidoptera: Noctuidae), by *Copidosoma bakeri* (Howard) (Hymenoptera: Encyrtidae) and effect on crop damage. Can Entomol 125:329–335

Camors FB, Payne TL 1972 Response of *Heydenia unica* (Hymenoptera: Pteromalidae) to *Dentroctonus frontalis* (Coleoptera: Scolytidae) pheromones and a host tree terpene. Ann Entomol Soc Am 65:31–33

De Moraes CM, Lewis WJ, Paré PW, Alborn HT, Tumlinson JH 1998 Herbivore-infested plants selectively attract parasitoids. Nature 393:570–573

Dicke M 1995 Why do plants 'talk'? Chemoecology 6:159–165

Dicke M 1999 Variation in induced plant–carnivore interactions in tritrophic systems. In: Insect–plant interactions and induced plant defence. Wiley, Chichester (Novartis Found Symp 223) p 43–59

Dicke M, van Beek TA, Posthumus MA, Ben Dom N, Van Bokhoven H, De Groot AE 1990a Isolation and identification of volatile kairomone that affects acarine predator–prey interactions. Involvement of host plant in its production. J Chem Ecol 16:381–396

Dicke M, Sabelis MW, Takabayashi J, Bruin J, Posthumus MA 1990b Plant strategies of manipulating predator–prey interactions through allelochemicals: prospects for application in pest control. J Chem Ecol 16:3091–3118

Elzen GW, Williams HJ, Vinson SB 1984 Isolation and identification of cotton synomones mediating searching behavior by parasitoid *Campoletis sonorensis*. J Chem Ecol 10:1251–1264

Jalali SK, Singh SP, Ballal CR 1988 Effect of parasitism by *Cotesia marginiventris* on consumption and utilization of artificial diet by larvae of *Spodoptera litura* (Lepidoptera: Noctuidae). Indian J Agric Sci 58:529–532

Karban R, Baldwin IT 1997 Induced responses to herbivory. University of Chicago Press, Chicago, IL

Lewis WJ, Tumlinson JH 1988 Host detection by chemically mediated associative learning in a parasitic wasp. Nature 331:257–259

Mattiacci L, Dicke M, Posthumus MA 1994 Induction of parasitoid attracting synomone in brussels sprouts plants by feeding of *Pieris brassicae* larvae: role of mechanical damage and herbivore elicitor. J Chem Ecol 20:2229–2247

Mattiacci L, Dicke M, Posthumus MA 1995 β-glucosidase: an elicitor of herbivore-induced plant odor that attracts host-searching parasitic wasps. Proc Natl Acad Sci USA 92:2036–2040

McCall PJ, Turlings TCJ, Lewis WJ, Tumlinson JH 1993 Role of plant volatiles in host location by the specialist parasitoid *Microplitis croceipes* Cresson (Braconidae: Hymenoptera). J Insect Behav 6:625–639

Parker FD, Pinnell RE 1973 Effect of food consumption of the imported cabbageworm when parasitized by two species of *Apanteles*. Environ Entomol 2:216–219

Potting RPJ, Vet LEM, Dicke M 1995 Host microhabitat location by the stemborer parasitoid *Cotesia flavipes*: the role of locally and systemically induced plant volatiles. J Chem Ecol 21: 525–539

Röse USR, Manukian A, Heath RR, Tumlinson JH 1996 Volatile semiochemicals released from undamaged cotton leaves: a systemic response of living plants to caterpillar damage. Plant Physiol 111:487–495

Sabelis MW, De Jong MCM 1988 Should all plants recruit bodyguards? Conditions for a polymorphic ESS of synomone production in plants. Oikos 53:247–252

Steinberg SM, Dicke M, Vet LEM 1993 Relative importance of infochemicals from first and second trophic level in long-range host location by the larval parasitoid *Cotesia glomerata*. J Chem Ecol 19:47–59

Takabayashi J, Takahashi S, Dicke M, Posthumus MA 1995 Developmental stage of herbivore *Pseudaletia separata* affects production of herbivore-induced synomone by corn plants. J Chem Ecol 3:273–287

Tumlinson JH, Paré PW, Lewis WJ 1999 Plant production of volatile semiochemicals in response to insect-derived elicitors. In: Insect–plant interactions and induced plant defence. Wiley, Chichester (Novartis Found Symp 223) p 95–109

Turlings TCJ, Tumlinson JH, Lewis WJ 1990 Exploitation of herbivore-induced plant odors by host-seeking wasps. Science 250:1251–1253

Turlings TCJ, Tumlinson JH, Eller FJ, Lewis WJ 1991a Larval-damaged plants: source of volatile synomones that guide the parasitoid *Cotesia marginiventris* to the micro-habitat of its hosts. Entomol Exp Appl 58:75–82

Turlings TCJ, Tumlinson JH, Heath RR, Proveaux AT, Doolittle RE 1991b Isolation and identification of allelochemicals that attract the larval parasitoid, *Cotesia marginiventris* (Cresson), to the microhabitat of one of its hosts. J Chem Ecol 17:2235–2251

Turlings TCJ, McCall PJ, Alborn HT, Tumlinson JH 1993a An elicitor in caterpillar oral secretions that induces corn seedlings to emit chemical signals attractive to parasitic wasps. J Chem Ecol 19:411–425

Turlings TCJ, Wäckers F, Vet LEM, Lewis WJ, Tumlinson JH 1993b Learning of host-finding cues by hymenopterous parasitoids. In: Papaj DR, Lewis A (eds) Insect learning: ecological and evolutionary perspectives. Chapman & Hall, New York, p 51–78

Turlings TCJ, Loughrin JH, McCall PJ, Röse U, Lewis WJ, Tumlinson JH 1995 How caterpillar-damaged plants protect themselves by attracting parasitic wasps. Proc Natl Acad Sci USA 92:4169–4174

Turlings TCJ, Lengwiler UB, Bernasconi ML, Wechsler D 1998 Timing of induced volatile emissions in maize seedlings. Planta 207:146–152

Udayagiri S, Jones RL 1992 Role of plant odor in parasitism of European corn borer by braconid specialist parasitoid *Macrocentrus grandii* Goidanich: isolation and characterization of plant synomones eliciting parasitoid flight response. J Chem Ecol 18:1841–1855

Vet LEM, Lewis WJ, Cardé RT 1995 Parasitoid foraging and learning. In: Cardé RT, Bell WJ (eds) Chemical ecology of insects 2. Chapman & Hall, New York, p 65–101

DISCUSSION

Boland: You used preparative GC to separate the volatiles into four groups. This technique may seriously affect the composition, at least in the high boiling components. Did you check that in the effluent the original ratio is maintained?

Turlings: It is difficult to re-collect some of the highly volatile compounds in the glass tube. However, we played with this and in the end we were able to re-collect more than 90% of the highest volatility compounds by using some Super-Q absorbent in the glass tube and submerging the tube in liquid nitrogen. Therefore I am confident that we were close to the original ratios. This doesn't necessarily mean that we got out what we injected in, because decomposition may take place. But judging by the response of the parasitoids to the re-collected volatiles, I'm sure that we got the attractive compounds.

Pickett: How did you eliminate the effects of variation in the volatility of the different compounds in the bioassay work? When you put the green leaf compounds on the filter paper, they will come off more quickly, whereas fraction 4 will come off slowly. The terpenoids are also held by the paper to some extent.

Turlings: We didn't. We did not do anything in this particular experiment to slow down the release rates of these compounds. If we did, we would possibly get a more natural release rate of the highly volatile compounds, but then a very unnatural release of the less volatile compounds. In this case I just applied the solvent with the volatiles directly onto the paper, waited just a minute to let the solvent evaporate and then I placed it in the wind tunnel. I used the same strip of paper throughout an experiment that may last 30–40 minutes. So there is likely a range of different release rates for the various compounds.

Gershenzon: The compounds may affect each other's release rate so that the volatility of a substance in a mixture is very often different from volatility when measured in pure form. Hence, it would be interesting to follow up to see what the headspace composition of these fractions really shows you, in comparison to the composition of what you have applied to the filter paper.

Turlings: If you look at those headspaces you get complicated results that depend on how long you collect for and precisely when you collect. It is far more useful to start working on a technique to get those release rates more comparable to the natural situation.

Gershenzon: Regarding your synthetic blend of volatiles, did you mix and test your blend on the same day, or mix it up on one day and test it a few days later? I'm curious about the stability of the individual compounds in these blends. Was the chromatogram that you showed us obtained from a sample taken at the same time that you conducted the bioassay?

Turlings: I don't remember how long a time separated that chromatogram and the bioassay. We used just one synthetic blend for the whole series of experiments, and that lasted about 3–4 weeks.

Pickett: Certainly the farnesene and the homoterpenes are unstable.

Turlings: They're not unstable if you look at them in terms of the natural blend. We do a lot of re-analysis of samples that we collected several months previously, and we get the same results. Admittedly, these are kept at $-70\,^\circ$C.

Gershenzon: This is interesting; they tend to be more stable in their natural context than when they are purified.

Pickett: Purified farnesenes oxidize very quickly.

Dietrich: Can you further fractionate fractions 3 and 4?

Turlings: Yes, that is something that would be interesting to look at in the future, but we haven't done it yet.

Beale: Fraction 2 as a whole had a negative effect on the bioactivity of 3 plus 4. Therefore, have you added back the individual components of 2?

Turlings: We haven't done that yet.

Vet: You are saying that fraction 2 changes the responsiveness to 3 and 4, and you are talking in terms of a deterrent effect. But might it instead be the case that fraction 2 is actually making the signal from the other fractions less clear? That is, if 3 and 4 are really the fractions that elicit the parasitoid behaviour and you add something else, the clarity of the signal will be reduced. Consequently, it misleadingly looks like fraction 2 has a deterrent effect.

Turlings: That's definitely a possibility. I think we have to get rid of that problem with the release rates and then we can start playing around with adding and leaving out certain compounds.

Tumlinson: This could be something to do with learning. These wasps are experienced at detecting the odour of caterpillars feeding on plants, and this signal depends on what the plants were releasing at the time. It's hard to say that some compounds are more important than others, because the plants release different blends at different times of the day.

Vet: So 1 and 2 may not have been there during the training and therefore the insects really go for 3 and 4. And when you train them on immediate plant response — basically on the C_6 compounds — you may get more of a response to fractions 1 and 2.

Turlings: I did an experiment that sort of addresses this issue, but I didn't do it for that purpose. We compared fresh damage with old damage, and tested those next to each other in the wind tunnel. We gave a group of wasps experience with fresh damage (mainly the green leafy volatiles released) and then another group of wasps had experience with the old damage (terpenoids). Wasps that have had experience with old damage show a much clearer preference for old damage, but the ones that had experience with fresh damage were still quite attracted to the old damage.

Poppy: I suppose another way of looking at this is asking whether you are dealing with recognition of the correct 'signal', or recognition of an incorrect 'signal'. With aphids, for instance, the parasitoid *Aphidius ervi* has an innate ability to discriminate between bean plants infested by its preferred host, *Acyrthosiphon pisum*, from bean plants infested by the non-host, *Aphis fabae*. What we don't know at this stage is whether that's because *A. fabae* induces the plant to produce an incorrect 'signal', which the parasitoids recognize and do not respond to, or because this aphid does not induce the plant to produce the correct 'signal'.

In terms of fractions 1 and 2, is it actually just a dilution factor or is there actually an incorrect signal that's been recognized so the parasitoids no longer respond to the correct signal? Are they so plastic that sometimes you can get more discrimination if that incorrect signal has been associated with a negative experience?

Baldwin: Since you have all demonstrated that associative learning is fundamentally important in this process, and in the light of this discussion of trying to understand the source of variation in volatile emissions from plants, I was wondering whether or not in your tests purified volicitin generates the same degree of variability that you see when you test the whole spit. You have talked about many different sources of variation: when you replace a herbivore with a defined chemical substance, does some of that variation go away or does the emission from the plant continue to be variable?

Tumlinson: Whether the seedlings are incubated in spit or in volicitin doesn't alter the variation in volatile release with time of day. Of course, volicitin isn't the only active compound in the spit: we know that there's at least one other compound that has activity. I'm not sure that testing pure volicitin is therefore going to give you all the answers. We see a lot of variability with the plants no matter what we do.

Haukioja: Insect spit contains lots of microorganisms: do you know whether these microorganisms produce the significant signals, or is it something from the insect itself?

Tumlinson: No.

Haukioja: Would it be possible to kill the microorganisms in the spit by irradiation, for example, so that you could test this?

Tumlinson: I don't think that would be the experiment. Once you have the spit the compounds are there, so killing the microorganisms at that point is not going to answer to the question.

Turlings: Killing microorganisms would definitely not reduce the activity of the spit: if you boil the spit you still have activity, and it is also active when it has been sterile-filtered. But as Jim Tumlinson said, this still doesn't rule out the possibility that the elicitor is already produced by microorganisms at this stage.

Schultz: At the risk of interrupting the terpenoid flow here, I'm curious about the indole. The question I've been wondering about for a long time is, if a plant is exposed to indole, does it have physiological activity? After all, indole is volatile and may be perceived by plants as a form of growth hormone or signal.

Firn: Indole tends to be pretty inactive in changing plant growth and development at realistic concentrations.

When plants are attacked by fungi, do they give off volatiles?

Pickett: Yes.

Firn: How do fungal-induced volatiles compare with the volatiles induced by an insect pest? How can a parasite discriminate between the signals?

Pickett: In the situation that we know best, which is *Alternaria brassicae* on oilseed rape (*Brassica napus*), the *A. brassicae* causes a different quantitative profile and the generation of some novel compounds. In fact, one is so novel we don't know what it is yet.

Firn: Is there a difference in parasitism when comparing plants carrying a fungal infection and healthy plants?

Pickett: This would most likely complicate the studies.

Dicke: The volatile profile released by lima bean plants infected with *Pseudomonas* bacteria is completely different from the profile induced by spider mite infection, as I will show in my paper (Dicke 1999, this volume).

Vet: Some parasitoids already have problems when two different herbivores are feeding on the same host plant.

Firn: The message emerging here is that the information content in volatile mixes is extremely high, especially when one considers mixed plant communities with multiple pests and diseases.

Dicke: That might be very relevant in the context of what Ted Turlings showed, because if an attacked plant is in the midst of plants that have not been attacked but which have still been damaged by wind, for instance, they will produce the green leaf volatiles but they don't produce the terpenoids. Thus there will be a plant that emits the complete mixture surrounded by plants that emit only part of the mixture.

Firn: Presumably the signals from any single plant species will be blended with signals from other species of plants producing different volatiles. The volatiles sensed by any parasite will therefore be made up of many compounds from a wide range of plants. This means that the signal will be extremely 'noisy', with high spatial and temporal variation.

Wadhams: On this issue of how insects can perceive a point source, Tom Baker has recently published an interesting paper on cell pairings in Lepidopteran male moths (Baker 1998). This focuses the orientation of the male to a point source. Because of a temporal coding that occurs, not only are they able to look at ratios, but they look at the ratios at an instantaneous point in time. This applies even more when you're talking about phytophagous insects, and host location and herbivores. Most herbivores are responding at a peripheral level to ubiquitous plant compounds, yet they're still able to differentiate their host from a non-host. The way we think they do that is by looking for this instantaneous stimulation of the two olfactory cells, which says that a particular signal comes from that plant and it's not fortuitous mixing of signals from two plants giving the same appropriate ratio.

Haukioja: As you mentioned in your paper, in some cases parasitized herbivore larvae actually consume more than unparasitized larvae. However, in the case you described it was clear that the parasitized larvae ate less of the plant. Do you have any idea of whether in this case reduced consumption by the larva after parasitization is perhaps specific for this herbivore or specific for this parasitoid? If there are differences, for instance, in how parasitoids manipulate larval feeding, for plants to really benefit, they should attract those parasitoids which cause the larva to eat less.

Turlings: This particular parasitoid, independent of which species it attacks, will reduce the feeding rate. Whether all caterpillars that are feeding on this particular plant are attacked by parasitoids that reduce feeding rates, I don't know. I don't think that it's up to the plant to send out a signal that only attracts this specific parasitoid and not others, because if the others are interested in finding the caterpillars they can pick up the same signal.

Vet: You are working with a solitary parasitoid. We have two species of parasitoid being attracted by the plants. One is the solitary parasitoid, and in general the solitary parasitoids have a limited amount of resources that they use from the host and they generally tend to reduce the growth after parasitization. The gregarious parasitoid, which is also attracted, can be quite different. It is known from polyembryonic parasitoids (which lay one egg in a host, but this then multiplies to as many as a few thousand individuals), that they can stimulate the feeding behaviour of the host by 50% or more, so these are terrible for the plant. With our gregarious parasitoid, the effect on the growth of the herbivore depends on the clutch size that is being laid inside the host: if there is only a small clutch, host growth (and thus feeding rate) is reduced to the point where it's still less than an unparasitized host. But if there are many eggs, this stimulates the host to grow. You can see how difficult it can be if you don't look at the combination of species that can use this plant information — the plant cannot choose who is going to make use of the signals it releases.

Gierl: In your profiles of the volatiles released by plants, I was surprised to see that in the absence of induction there was absolutely no signal. How do the herbivores find their host plants in the first place?

Turlings: As far as our detection limits are concerned, it depends on which genotype of plant you look at. The example that I showed is fairly extreme: we really don't see anything. But in other cases linalool and α-farnesene are produced by undamaged plants, sometimes in quite significant amounts.

Gierl: And are these signals overlapping signals that are then emitted to attract the parasitoids?

Turlings: Yes.

Tumlinson: Undamaged plants do release volatiles, but they release them at very different levels. Typically, in some of our experiments with a cotton plant that had

6–8 true leaves, when it was damaged by several larvae we were getting 8–10 μg/h, whereas from an undamaged plant we were getting maybe 400 ng/h.

So many of our experiments are artificial: we take these plants and take great care not to damage them, and we put them in our volatile collection systems. In fact if you get a little bit of damage, the compounds that we always see that indicate stress just jump at you. I suspect that in the natural situation, due to wind damage and so forth and so on, you are going to get a lot of compounds released from the plants.

Wadhams: I agree. Intact plants give off almost no volatiles, but the insects can still detect those levels. In coupled GC/electrophysiological runs where we are using the insect's antennae as a detector, we can, in some instances, have an absolutely flat GC baseline but the insect's olfactory cells still respond strongly. There can be several orders of magnitude difference between the insect's ability to detect a compound and our ability to see it with current techniques.

References

Baker TC 1998 Species specificity of pheromone responses: receptor neurons mediate behavioural responses in heliothine moths. The Biochemist 20(4):26–29

Dicke M 1999 Specificity of herbivore-induced plant defences. In: Insect–plant interactions and induced plant defence. Wiley, Chichester (Novartis Found Symp 223) p 43–59

General discussion I

Poppy: Our discussions of the co-evolutionary aspects of herbivore-induced volatiles in tritrophic interactions, and how plants can benefit if the larval consumption rate is reduced, still don't answer the key question of what is actually happening within the plant. Is the plant actually generating this 'signal' as a response to herbivore attack and therefore those plants that do it well are selected over evolutionary time? Or, as we believe is the case for aphids, is the herbivore saliva actually causing up- and down-regulation of various biosynthetic pathways within the plant to make that plant more suitable to the herbivore, and the parasitoids are exploiting volatiles which are produced as a consequence of this up- and down-regulation? We are still not sure which is the governing factor: are the plants actively responding and sending out SOS signals, or are the volatiles produced as a by-product of being attacked?

Baldwin: In thinking about the function of these signals, when you are devoid of information it is sometimes helpful to approach this question from the engineering standpoint, even though that's not the way evolution works (evolution clearly functions as a tinker rather than an engineer). Richard Karban came up with an alternative idea about how induced defences function—his 'moving target' model (Adler & Karban 1994). In this model he argued that what was important functionally for an induced response was not that it was a directional response towards a phenotypic state that was necessarily more defended, but the change in phenotypic state was a way to generate variation. And it is the variance *per se* that functions defensively. When you're dealing with an information-based defence, which as Louise Vet has pointed out so many times, is not controlled by the plant—it is just thrown out and the rest of the ecological community responds in a highly complex way—there may be a lot of selection for generating variation in that signal. I wonder whether when we are designing analytical procedures to measure these responses, we are actually throwing away part of the functional component.

Vet: I think variation may mean something different for a herbivore than for their natural enemies. Learning seems to play an essential role for natural enemies. It may be that plants can indeed disturb the attraction of the herbivore by creating variation. And because the natural enemies are such good learners, they may pick up the essential variation and the signal may therefore still be of benefit to the plant. The plants may be able to exploit the differential response of herbivores and their natural enemies.

Firn: The fact that many volatiles are inducible by mechanical damage, herbivory or infection should not necessarily be regarded as remarkable. If organisms gain fitness by possessing mechanisms to generate and retain chemical diversity (Jones & Firn 1991), then it is predictable that mechanisms that give cost reductions will be selected.

One such mechanism would be inducibility. The linking of the control elements of the biosynthetic pathway with an appropriate sensor to give inducibility could occur early in the evolutionary history of the pathway. The gains from subsequent selection improving this inducibility of the overall pathway might be expected to exceed the gains of making each later stage of the pathway inducible. It might be expected that more than one sensor would be linked to the inducible elements such that different cues might all enhance the flux through the pathway, with different individual compounds in the resulting mixture conferring a benefit under different circumstances. This would explain why we have heard of so many stimuli that can induce volatile production (mechanical damage, insects, bacteria, fungi, etc.) and why hundreds of different compounds might be induced by any one stimulus.

Schultz: The variance idea is a more useful explanation for variation in resistance traits by plants that act on the herbivore. I think there's value in being undependable with respect to the herbivore. It is essentially a resistance management strategy: you make the resistance variable so the herbivore has trouble tracking it. In this case we want the parasitoid to track this variation.

Baldwin: But you cannot guarantee that only the parasitoid picks up on it. As long as you can't guarantee that, and you've got an organism (the parasitoid) which is potentially a quicker learner than another organism (the herbivore), perhaps changing what a plant emits is a way to make sure that it's only the parasitoid that picks it up.

Schultz: That's true, but the evidence so far is that the suite of plant compounds that herbivores make use of isn't the same as the suite used by the parasitoids.

Vet: Do we know that for sure?

Städler: There are few examples. As far as I am aware, those examples that we do have indicate that this long-range orientation of herbivores to host plants does not exist. It is very different from the parasitoid situation. I would speculate that host plants that put out such compounds would be wiped out pretty soon.

Dicke: But there are very few examples where it has been shown that parasitoids are attracted over a range of 50–100 m. Most of these studies have been done in the lab — there are hardly any field data on this.

Turlings: As far as the few examples of herbivores responding to induced volatiles are concerned, the only cases where you find real attractiveness is when you look at beetles. Other examples show more repellency. We have shown that aphids are repelled by induced volatiles of maize plants.

Dicke: Peter Landolt (1993) showed that there is attraction of cabbage looper moths towards plants infested by conspecific larvae, but when they oviposit, they will oviposit on the plants that are close by, but are not actually infested by herbivores. So that is still an initial attraction that is used to locate suitable host plants.

Turlings: He found for cotton that infested plants were attractive to the moth *Trichoplusia ni*, but it preferred to oviposit on undamaged plants. For cabbage he found that healthy plants were more attractive than plants damaged by conspecifics. In both cases the moth avoids ovipositing on plants that are already infested. I would call that repellency.

Tumlinson: We shouldn't forget the bark beetle situation, where attack on the tree by the beetles induces the tree to produce volatiles that attract large numbers of beetles.

Schultz: Do these volatiles overlap with parasitoid attractants?

Tumlinson: Yes, parasitoids are also attracted.

Dicke: I was surprised, because in the 1980s I published on the response of spider mites towards volatiles from spider mite-infested plants (Dicke 1986). In this case the spider mites were repelled, and I thought that would prove to be a more general result. But now I can come up with more examples where herbivores are attracted than where herbivores are repelled by infested plants (see Dicke & Vet 1998). It might have to do with the herbivore density and consequently the amounts of volatiles that are emitted anyway, as was the case in my study. When the amount of volatiles increases, it is a great risk to the plant because it exposes itself and so herbivores then could exploit these volatiles that are produced.

Pickett: In our studies, we have both situations. Some of the infestation signals are used as attractants, admittedly under artificial circumstances, and some are used as repellents. However, some of the repellents can be highly effective in the field.

Wadhams: The sort of compounds that we are looking at, as far as I can see, are very similar, I just don't have a good documented case of a parasitoid and a herbivore where we know that they are looking for exactly the same compounds.

Pickett: Collectively, I think we ought to look more at this. We have excellent model systems; we have simply not yet done the work. With the parasitoid *Aphidius ervi* and the pea aphid *Acyrthosiphon pisum* we can look very carefully to see exactly which compounds are involved.

Firn: Are there parasites of parasites?

Pickett: Yes. There exists a complicated guild of hyperparasitoids.

Firn: Do these parasites use plant volatiles to detect their hosts?

Wadhams: There's one compound in the bean aphid/parasitoid complex, which we think is particularly associated with the parasitoid recognizing the aphid/host plant complex, and this may also be a pheromone for the hyperparasitoid.

Vet: Perhaps there is a reason why you can expect herbivores to respond differently compared to their natural enemies, because herbivores tend to have just one host plant species. One trophic level up, parasitoids generally have more than one host. This by itself expands the plant range immediately. Now if you assume that a generalist insect needs to learn more than a specialist, you would expect that herbivores will use more specific plant cues than the parasitoids one trophic level up.

References

Adler FR, Karban R 1994 Defended fortresses or moving targets? Another model of inducible defenses inspired by military metaphors. Am Nat 144:813–822

Dicke M 1986 Volatile spider-mite pheromone and host-plant kairomone, involved in spaced-out gregariousness in the spider mite *Tetranychus urticae*. Physiol Entomol 11:251–262

Dicke M, Vet LEM 1998 Plant–carnivore interactions: evolutionary and ecological consequences for plant, herbivore and carnivore. In: Olff H, Brown VK, Drent RH (eds) Herbivores: between plants and predators. Blackwell Science, Oxford, p 483–520

Jones CG, Firn RD 1991 On the evolution of secondary plant chemical diversity. Philos Trans R Soc Lond B Biol Sci 333:273–280

Landolt PJ 1993 Effects of host plant leaf damage on cabbage looper moth attraction and oviposition. Entomol Exp Appl 67:79–85

Specificity of herbivore-induced plant defences

Marcel Dicke

Laboratory of Entomology, Wageningen Agricultural University, P.O. Box 8031, 6700 EH Wageningen, The Netherlands

Abstract. In addition to induced direct defence, plants can defend themselves indirectly by improving the effectiveness of enemies of herbivores. Plants can respond to arthropod herbivory with the induction of a blend of volatiles that attracts predators and/or parasitoids of herbivores. Carnivorous arthropods can discriminate between infested plants and mechanically wounded plants, and between plants infested by different herbivore species. The volatile blends emitted by different plant species infested by the same herbivore species show large qualitative differences, whereas blends emitted by plants of the same species, but infested by different herbivore species are mostly qualitatively similar with quantitative variation. Carnivores can discriminate between blends that differ qualitatively and/or quantitatively. However, it remains unknown what differences in blends are used by carnivorous arthropods in this discrimination. Signal transduction pathways involved in the induction of direct and indirect defence seem to overlap. Direct and indirect defence may interfere with each other's effectiveness. For application of direct and indirect defence in agriculture, it is important to compare the relative importance of these two defence types in the same plant species.

1999 Insect–plant interactions and induced plant defence. Wiley, Chichester (Novartis Foundation Symposium 223) p 43–59

Insects dominate the world's fauna in terms of number of species and individuals. Plants are under the constant threat of being attacked by insects. They have evolved a fascinating array of defences that can be grouped into two categories: (1) direct defence that affects the attackers directly through e.g. toxins, repellents, digestibility reducers, spines and thorns; and (2) indirect defence that promotes the effectiveness of natural enemies of herbivores through e.g. the provision of shelter, alternative food or the production of SOS signals that enable carnivorous arthropods to locate the herbivores. Each of these two defences can be constitutively present or can be induced by attack.

This chapter will concentrate on induced defences and especially on the relative contribution of induced direct and induced indirect defences. The induction of

43

defences may occur in response to mechanical damage and/or herbivore damage. Whether a plant responds differently to mechanical damage vs. herbivore damage or to different herbivore species is a measure of the specificity of the plant's induced response. Different degrees of specificity of the plant's response to damage can be distinguished, depending on quantitative and qualitative differences in the responses to mechanical damage, and damage inflicted by different herbivore species. Qualitative differences relate to the production of different compounds in different treatments, while quantitative differences relate to the production of the same set of compounds but in different relative intensities. Qualitative differences in plant response characterize a higher degree of specificity than quantitative differences.

Induced direct defence

Induced direct defence against herbivores has been recorded since the 1970s for more than 100 plant species in 34 families (Karban & Baldwin 1997). Induced direct defence can comprise the induction of various characteristics such as toxins, digestibility reducers, repellents and trichomes (Karban & Baldwin 1997). Several plant species, such as tomato, potato, tobacco, and cotton have been studied in considerable depth.

Specificity of induced direct defence: chemical and biological evidence

In many cases direct defence can be induced by mechanically wounding a plant or by herbivore feeding damage, often with similar effects. For instance, mechanical wounding just like herbivore damage results in the induction of proteinase inhibitors in tomato and potato plants, leading to a reduction of the digestibility of plant tissues for herbivores (Green & Ryan 1971). Mechanical wounding and caterpillar (*Manduca sexta*) damage both induce the production of nicotine in tobacco plants, albeit that caterpillar damage results in lower levels of nicotine than mechanical damage (McCloud & Baldwin 1997). However, some reports demonstrate that induced plant responses can be specific for the type of damage that is inflicted. For instance, caterpillars (*Helicoverpa zea*), leafminer flies (*Liriomyza trifolii*) and russet mites (*Aculops lycopersici*) induce different combinations of polyphenol oxidase, peroxidase, lipoxygenase and proteinase inhibitors in tomato plants (Stout et al 1994).

When studying the effect on herbivores, specificity may be more difficult to detect. Different plant responses may have a similar effect on herbivores because each combination of induced responses negatively affects subsequent herbivores attacking the plant. In fact, the induction of toxins and more likely that of digestibility reducers will affect a range of insects. This is similar to the observation that constitutively present secondary plant chemicals have similar effects on a variety of herbivore species, with only specialist species being able to overcome the effects of the phytochemicals. For example, spider mite damage to cotton plants results in lower population increase during subsequent spider mite infestations (Karban & Carey 1984). Similar results are obtained with mechanical wounding or fungus (*Verticillium dahliae*) infestation (Karban et al 1987).

Induced indirect defence

The major form of induced indirect defence is the emission of herbivore-induced plant volatiles that attract carnivorous enemies of the herbivores. This induced defence has been reported since the 1980s for more than 20 plant species in 13 families (see review by Dicke 1999). Two major categories of plant response may be distinguished (Dicke 1999). (1) The emission of large amounts of novel compounds that dominate the blend from herbivore-infested plants. These novel compounds are not emitted, or only emitted in trace amounts, by undamaged plants or by mechanically wounded plants. Examples comprise lima bean, corn, and cucumber. (2) Plants that emit qualitatively similar but quantitatively different blends in response to herbivory or mechanical wounding or when undamaged. Examples include tomato, potato and cabbage. A large amount of information is present on the effect of these induced plant volatiles on the behaviour of carnivorous arthropods such as predators and parasitoids. Herbivore-induced plant volatiles are very important to carnivorous arthropods. Their herbivore victims usually do not emit large amounts of volatiles: they are small components in the environment, and the emission of volatiles that attract their enemies is strongly selected against. Volatiles from undamaged plants, though more abundant than herbivore volatiles, do not provide information to carnivores on the presence of herbivores. In contrast, plant volatiles induced by herbivory are both easily detectable and often reliable indicators of not only herbivore presence but also herbivore identity. Thus, herbivore-induced plant volatiles provide a solution to the reliability–detectability problem that carnivorous arthropods face and consequently the production of herbivore-induced plant volatiles is an important mechanism of increasing the effectiveness of carnivorous enemies of herbivores (Vet & Dicke 1992).

Specificity of herbivore-induced plant volatiles: chemical and behavioural evidence

Induced carnivore attraction can occur rather specifically. This is obvious from chemical as well as behavioural evidence. Many plant species emit volatile blends that are qualitatively very different in response to mechanical or herbivore damage (Dicke 1999). For instance, lima bean leaves infested with spider mites emit a blend that is qualitatively very different from that emitted from mechanically wounded lima bean leaves (Dicke et al 1990). The predatory thrips *Scolothrips takahashii* discriminates among volatiles from spider-mite-infested lima bean leaves and mechanically wounded lima bean leaves; the thrips are not attracted by mechanically wounded leaves (Shimoda et al 1997). However, even when qualitative differences are absent or minor, carnivores may discriminate between mechanically damaged and herbivore-damaged plants. Cabbage leaves infested with *Pieris brassicae* caterpillars emit a qualitatively similar blend as mechanically wounded cabbage leaves, but caterpillar-infested leaves emit larger amounts and for a longer time period after damaging the plant halts (Mattiacci et al 1994). The parasitoid *Cotesia glomerata* is attracted by mechanically wounded cabbage leaves, but when offered against cabbage leaves infested with their hosts, *P. brassicae* caterpillars, they prefer the volatiles from caterpillar-infested cabbage leaves (Steinberg et al 1993). Plants can also emit different volatile blends in response to pathogens or herbivores. For instance, bean plants (*Phaseolus vulgaris* cv. Red Mexican) that are infested with the bacterium *Pseudomonas syringae* pv. *phaseolica* emit a blend of volatiles that is produced through the lipoxygenase pathway, with (Z)-3-hexen-1-ol as major component, but no terpenoids are emitted (Croft et al 1993, A. J. Slusarenko, personal communication). In contrast, when bean plants of the same cultivar are infested with the spider mite *Tetranychus urticae*, the plants emit a blend in which (Z)-3-hexen-1-ol is a minor component, several terpenoids are induced and the homoterpene 4,8,-dimethyl-1,3(E),7-nonatriene is a dominant blend component (M. Dicke & M. A. Posthumus, unpublished results; Fig. 1).

In addition to differences in response to mechanical wounding, pathogen infestation and herbivory, plants may also emit different blends when infested by different herbivore species. For instance, apple foliage infested with either of two herbivorous mite species, *Panonychus ulmi* or *T. urticae*, emit blends that, although qualitatively similar, differ in the relative contribution of the compounds to the total blend (Takabayashi et al 1991a). Carnivorous mites discriminate between volatiles emitted by apple plants infested by either of the two herbivorous mite species. The predators *Amblyseius finlandicus* and *A. andersoni* are attracted to apple foliage infested with *P. ulmi*, their preferred prey species and not to apple foliage infested with *T. urticae*, while the carnivore *Phytoseiulus persimilis*, a specialist predator of *T. urticae*, has the reverse response (Sabelis & Dicke 1985). The recorded quantitative differences in blend composition may allow the predatory mites to discriminate. However, it cannot be excluded that qualitative differences in components that were present below the detection level of the equipment, play a role. In some systems, chemical analyses reveal hardly any

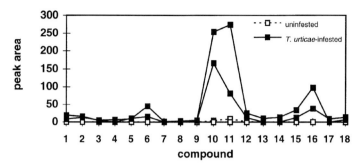

FIG. 1.   Headspace composition of uninfested or spider-mite-infested (*Tetranychus urticae*) bean plants (*Phaseolus vulgaris*, cultivar Red Mexican). Total peak areas (arbitrary units) recorded for spider-mite-infested plants were 427 and 834, and for undamaged plants 65 and 0. Compound numbers: 1, 2-butanone; 2, 3-pentanone; 3, 2-methylpropanenitrile; 4, 1-penten-3-one; 5, 2-butanol; 6, 1-penten-3-ol; 7, limonene; 8, (*E*)-2-hexenal; 9, (*E*)-*β*-ocimene; 10, (*E*)-4,8-dimethyl-1,3,7-nonatriene; 11, (*Z*)-3-hexen-1-ol acetate; 12, (*Z*)-3-hexen-1-ol; 13, 2-methylpropanal oxime; 14, 1-octen-3-ol; 15, methyl salicylate; 16, (*E,E*)-4,8,12-trimethyl-1,3,7,11-tetradecaene; 17, benzyl alcohol; 18, indole.

differences in blend composition, either in qualitative or in quantitative respect, while carnivorous enemies of the herbivores are able to discriminate among the blends. An example is the discrimination by the parasitoid *C. glomerata* between cabbage plants infested by either caterpillars of *P. brassicae* or caterpillars of *Pieris rapae* (Geervliet et al 1998). Chemical analyses revealed no qualitative differences and only minor quantitative differences among the blends of cabbage plants infested by either caterpillar species (Blaakmeer et al 1994, Geervliet et al 1997). The discrimination by the parasitoids was only recorded after several oviposition experiences (Geervliet et al 1998). Learning to discriminate among very similar odour blends has been recorded for other carnivorous arthropods as well (Turlings et al 1993, Vet et al 1998, Dicke & Vet 1999). Moreover, a few examples are known where parasitoids that had no previous oviposition experience discriminated between odour blends emitted by plants infested with different herbivores. In these instances, the odour blends differed qualitatively (Takabayashi et al 1995, Powell et al 1998). The ability of carnivores to discriminate between plants infested by different herbivore species has been recorded for a range of plant–herbivore–carnivore systems (see Dicke & Vet 1999 for review).

What variation in blend composition is important to carnivorous arthropods?

Initially, odour blends emitted by herbivore-infested plants seemed to have a simple composition, comprising 10–20 compounds (Dicke et al 1990, Turlings et al 1990). However, with the development of more sensitive analytical methods, more and more blend components have been identified. For instance, in 1990 we reported 17 compounds in the blend emitted by spider-mite-infested lima bean plants (Dicke et al 1990). Recent analyses resulted in the identification of more than 60 compounds from the same odour blend (M. Dicke, R. Gols, D. Ludeking & M. A. Posthumus, unpublished data). Odour blends emitted by herbivore-infested plants may be composed of 100 or more compounds, many of which occur as minor constituents (R. Gols, O. E. Krips, M. A. Posthumus & M. Dicke, unpublished data). The composition of the volatile blends induced by herbivory varies with plant species, plant genotype, leaf age, time of the day and herbivore species or instar that damages the plant (Turlings et al 1995, Takabayashi & Dicke 1996). Among plant species, blend composition varies qualitatively in many components although shared compounds also occur. Commonly induced compounds are e.g. the two homoterpenes 4,8,-dimethyl-1,3(E),7-nonatriene and 4,8,12-trimethyl-1,3(E),7(E),11-tridecatetraene (Dicke et al 1990, Turlings et al 1990, 1995, Takabayashi & Dicke 1996). In contrast, among plants of the same species, blend composition varies predominantly quantitatively, leading to blends that differ in ratios of similar compounds (Takabayashi et al 1991a, Turlings et al 1993, De Moraes et al 1998).

Although a lot of knowledge exists on the chemical compositions of herbivore-induced plant odour blends and on the behavioural responses of carnivorous arthropods to the blends emitted by herbivore-damaged plants, little is known about the relative importance of blend components for the response of carnivores. This is partly due to the complex blend composition, which sharply contrasts to e.g. the composition of sex pheromones of moths. For instance, in studies on sex pheromones, the total number of compounds emitted from the pheromone gland is usually restricted to 2–10 (Cardé & Minks 1997), while infested plants can emit tens to more than 100 different compounds (Turlings et al 1993, Mattiacci et al 1994, R. Gols, O. E. Krips, M. A. Posthumus & M. Dicke, unpublished data). The parasitoid *Cotesia marginiventris* is attracted to a synthetic mixture of the 11 major components of the blend emitted by corn plants infested by their hosts, *Spodoptera exigua*. However, although the synthetic mixture was composed as a close mimic of the natural mixture, the parasitoids' response to the natural and synthetic mixture significantly differed (Turlings et al 1991). This may have been due to minor components in the natural blend or to small differences in the 11 major compounds between the natural blend and the synthetic blend. The predatory mite *P. persimilis* is attracted to four chemicals identified in the blend emitted by lima bean plants infested with their prey, the spider mite *T. urticae*, even though the compounds were offered individually (Dicke et al 1990). No comparison of the effect of the single components or a synthetic mixture of them and a natural odour blend has been made for this predatory mite. A behavioural response to six individual compounds identified in the mixture emitted by faba bean plants infested by the aphid *Acyrthosiphon pisum* was recorded in females of the parasitoid *Aphidius ervi*. Oviposition experience increased the attraction to each of the blend components (Du et al 1998).

The ability of carnivorous arthropods to discriminate among odour blends is significantly characterized by phenotypic plasticity. Important factors affecting behavioural responses include starvation, specific hunger, and successful and unsuccessful foraging experiences (Vet & Dicke 1992, Turlings et al 1993, Dicke et al 1998, Vet et al 1998, Vet 1999, this volume). For instance, the parasitoid *C. marginiventris* was strongly attracted to a synthetic odour blend made to mimic the natural, plant-emitted odour blend after the wasps had a successful oviposition experience in the presence of the synthetic blend, but not after a successful experience in the presence of the natural odour blend (Turlings et al 1991). Very subtle differences between odour blends may be learned by parasitoids and qualitative differences in odour blends can be learned with more ease than quantitative differences (Vet et al 1998, Vet 1999, this volume).

Some knowledge of the importance of variation in the composition of plant volatile blends for the behavioural response of arthropods is available for herbivorous beetles. The Colorado potato beetle is strongly attracted by volatiles emitted by potato foliage, either when undamaged, mechanically wounded, or damaged by herbivores (Visser & Ave 1978, Bolter et al 1997). Volatiles emitted by potato plants are not specific for potato, but are emitted by a wide variety of plant species. However, the composition of the *blend* is characteristic for potato. When individual components were added to the natural blend of potato so that quantitative variation in blend components was achieved, the behavioural response of the beetles was disturbed in several, but not all, treatments (Visser & Ave 1978).

Elucidating how carnivores cope with variation in blends of herbivore-induced plant volatiles will be an exciting challenge, now that chemical information on variation in blend composition and the factors influencing behavioural responses of carnivores are well known.

Specificity of SOS signals: how do plants do it?

The differences in plant responses to herbivory and mechanical wounding can be explained by the involvement of herbivore oral secretions. The application of oral secretion onto a mechanical wound results in the same response as herbivory (Turlings et al 1990, Mattiacci et al 1994, 1995, Alborn et al 1997). Oral secretions may also be applied through the plant's cut stem and then induce odour emission from undamaged plant leaves (Turlings et al 1995). Oral secretions may also explain the specificity of the plant's response to different herbivores. For instance, corn plants infested with 1st/2nd instar caterpillars of *Pseudaletia separata* emit a different volatile blend than corn plants infested with 5th/6th instars. Plants infested with 1st/2nd instars attract the parasitoid *Cotesia kariyai* whereas plants infested with 5th/6th instars do not. When oral secretions from young and old caterpillars were applied onto a mechanical wound a similar difference in response of the parasitoids was recorded (Takabayashi et al 1995). Interestingly, a single component of a herbivore's oral secretion can induce a complex volatile blend that is similar to the blend induced by herbivory (Mattiacci et al 1995, Alborn et al 1997). The blend components result from several different biosynthetic pathways such as the isoprenoid, shikimic acid and lipoxygenase pathways. It remains unclear how a single herbivore elicitor induces different pathways resulting in quantitative differences in blend composition. Herbivory (Blechert et al 1995) and oral secretion (McCloud & Baldwin 1997) can induce jasmonic acid in plants and jasmonic acid application to lima bean or corn plants results in the induction of e.g. the homoterpene 4,8,-dimethyl-1,3(E),7-nonatriene, which is also induced by herbivory (Hopke et al 1994). However, it seems that jasmonic acid cannot fully explain the plant's response. For instance, the homoterpene 4,8,12-trimethyl-1,3(E),7(E),11-tridecatetraene is not induced in bean plants by jasmonic acid application (Hopke et al 1994, M. Dicke, R. Gols, D. Ludeking & M. A. Posthumus, unpublished data). Also, in induced direct defence, jasmonic acid seems unable to explain the plant response completely (McCloud & Baldwin 1997). Induced indirect defence is expressed systemically (Takabayashi et al 1991b, Turlings & Tumlinson 1992, Dicke et al 1993) and a systemic elicitor can be extracted from plants (Dicke et al 1993). This systemic elicitor remains unidentified. In systemic induction of nicotine production in tobacco plants jasmonic acid is involved and this may be the systemically transported elicitor (McCloud & Baldwin 1997).

Are direct and indirect defence similarly induced?

Signal transduction pathways of induced direct and indirect defence seem to overlap. In both types of induced defence, the octadecanoid pathway through jasmonic acid seems to be central. However, the fact that different herbivores or their oral secretions can lead to different volatile blends also indicates that there are likely to be regulatory mechanisms that function differently in the two types of induced defence.

A problem for a sound comparison of signal transduction in induced direct and induced indirect defence, though, is that signal transduction in induced indirect defence is mainly studied in two plant species, bean and corn, that are characterized by differential responses to mechanical and herbivore damage in terms of induced volatiles. On the other hand, signal transduction in induced direct defence is studied mostly in plants that have similar responses, in terms of induced volatiles, to these two different types of damage: tomato, potato and cabbage. For an appropriate comparison of signal transduction in induced direct and indirect defence it is important that future studies combine an investigation of each of these defence types in a single plant species. Possibly, some plant species are characterized by specific responses to different types of damage, while other plant species are characterized by non-specific responses.

Relative importance of direct defence vs. indirect defence

The two types of defence, direct and indirect defence, available to plants may interfere with each other's effectiveness. For instance, tomato plants have an important direct defence in the form of glandular trichomes. These trichomes immobilize and intoxicate herbivores (van Haren et al 1987). However, these trichomes also negatively affect carnivorous arthropods, resulting in high carnivore mortality rates and thus in an impairment of indirect defence (van Haren et al 1987). Consequently, if tomato plants were to attract enemies of herbivores, the carnivores would die on the plant rather than feed on the herbivores. It would be interesting to investigate whether the intensities of direct and indirect defence are negatively correlated. Some observations seem to support this. For instance, the emission of a herbivore-induced volatile blend that is similar to the blend induced by wounding seems more primitive than the emission of a blend that is dominated by novel compounds. Plants that do not emit novel dominant blend components are well characterized by particular secondary metabolites that function as strong direct defences. Solanaceous plants such as tomato and potato, and crucifers such as cabbage are well known producers of alkaloids or glucosinolates, respectively, that confer strong resistance against many insect species. These plants emit similar blends in response to herbivory and mechanical wounding (Mattiacci et al 1994, Bolter et al 1997, Dicke et al 1998). They are often used as food only by a limited number of specialist herbivores such as the Colorado potato beetle or cabbage whites. In contrast, plants that emit novel compounds that dominate the induced blend are plants such as corn and bean. These plants do not have such highly specialized secondary metabolites and can be used as food by various herbivore species. For a conclusive answer to the question of whether the development of direct and indirect defences are negatively correlated, a comparison of the defence characteristics of a range of plant species is necessary.

Crop protection: combining direct and indirect defences

The question of the relative importance of direct and indirect defences is important for applications in agriculture. Direct defence can be applied in host plant resistance and indirect defence in biological control through predators or parasitoids. Pest management strategies have mostly been developed either through host plant resistance or through biological control. However, with a reduction of pesticide use the two methods of pest control will increasingly be applied in the same crop. Therefore, developments in either of the two pest control methods should take into consideration what the effects are on the other method so as to reach a synergistic set of control methods rather than a net reduction in pest control effectiveness.

Acknowledgements

I thank Louise E. M. Vet for constructive comments on a previous version of the manuscript. The research described is partially funded by the Uyttenboogaart-Eliasen Foundation.

References

Alborn T, Turlings TCJ, Jones TH, Steinhagen G, Loughrin JH, Tumlinson JH 1997 An elicitor of plant volatiles from beet armyworm oral secretion. Science 276:945–949

Blaakmeer A, Geervliet JBF, van Loon JJA, Posthumus MA, van Beek TA, De Groot AE 1994 Comparative headspace analysis of cabbage plants damaged by two species of *Pieris* caterpillars: consequences for in-flight host location by *Cotesia* parasitoids. Entomol Exp Appl 73:175–182

Blechert S, Brodschelm W, Holder S et al 1995 The octadecanoic pathway: signal molecules for the regulation of secondary pathways. Proc Natl Acad Sci USA 92:4099–4105

Bolter CJ, Dicke M, van Loon JJA, Visser JH, Posthumus MA 1997 Attraction of Colorado potato beetle to herbivore damaged plants during herbivory and after its termination. J Chem Ecol 23: 1003–1023

Cardé RT, Minks AK (eds) 1997 Insect pheromone research: new directions. Chapman & Hall, New York

Croft KP, Juttner F, Slusarenko AJ 1993 Volatile products of the lipoxygenase pathway evolved from *Phaseolus vulgaris* (L) leaves inoculated with *Pseudomonas syringae* pv *phaseolicola*. Plant Physiol 101:13–24

De Moraes CM, Lewis WJ, Paré PW, Alborn HT, Tumlinson JH 1998 Herbivore-infested plants selectively attract parasitoids. Nature 393:570–573

Dicke M 1999 Evolution of induced indirect defence of plants. In: Tollrian R, Harvell CD (eds) The ecology and evolution of inducible defenses. Princeton University Press, Princeton, NJ, p 62–88

Dicke M, Vet LEM 1999 Plant–carnivore interactions: evolutionary and ecological consequences for plant, herbivore and carnivore. In: Olff H, Brown VK, Drent RH (eds) Herbivores: between plants and predators. Blackwell Science, Oxford, p 483–520

Dicke M, van Beek TA, Posthumus MA, Ben Dom N, Van Bokhoven H, De Groot AE 1990 Isolation and identification of volatile kairomone that affects acarine predator–prey interactions. Involvement of host plant in its production. J Chem Ecol 16:381–396

Dicke M, van Baarlen P, Wessels R, Dijkman H 1993 Herbivory induces systemic production of plant volatiles that attract predators of the herbivore: extraction of endogenous elicitor. J Chem Ecol 19:581–599

Dicke M, Takabayashi J, Posthumus MA, Schutte C, Krips OE 1998 Plant–phytoseiid interactions mediated by herbivore-induced plant volatiles: variation in production of cues and in responses of predatory mites. Exp Appl Acarol 22:311–333

Du Y, Poppy GM, Powell W, Pickett JA, Wadhams LJ, Woodcock CM 1998 Identification of semiochemicals released during aphid feeding that attract the parasitoid *Aphidius ervi*. J Chem Ecol 24:1355–1368

Geervliet JBF, Posthumus MA, Vet LEM, Dicke M 1997 Comparative analysis of headspace volatiles from different caterpillar-infested and uninfested food plants of *Pieris* species. J Chem Ecol 23:2935–2954

Geervliet JBF, Vreugdenhil AI, Vet LEM, Dicke M 1998 Learning to discriminate between infochemicals from different plant–host complexes by the parasitoids *Cotesia glomerata* and *Cotesia rubecula* (Hymenoptera: Braconidae). Entomol Exp Appl 86:241–252

Green TR, Ryan CA 1971 Wound-induced proteinase inhibitor in plant leaves: a possible defence against insects. Science 175:776–777

Hopke J, Donath J, Blechert S, Boland W 1994 Herbivore-induced volatiles: the emission of acyclic homoterpenes from leaves of *Phaseolus lunatus* and *Zea mays* can be triggered by a β-glucosidase and jasmonic acid. FEBS Lett 352:146–150

Karban R, Baldwin IT 1997 Induced responses to herbivory. University of Chicago Press, Chicago, IL

Karban R, Carey JR 1984 Induced resistance of cotton seedlings to mites. Science 225:53–54

Karban R, Adamchak R, Schnathorst WC 1987 Induced resistance and interspecific competition between spider mites and a vascular wilt fungus. Science 235:678–680

Mattiacci L, Dicke M, Posthumus MA 1994 Induction of parasitoid attracting synomone in brussels sprouts plants by feeding of *Pieris brassicae* larvae: role of mechanical damage and herbivore elicitor. J Chem Ecol 20:2229–2247

Mattiacci L, Dicke M, Posthumus MA 1995 β-glucosidase: an elicitor of herbivore-induced plant odor that attracts host-searching parasitic wasps. Proc Natl Acad Sci USA 92:2036–2040

McCloud ES, Baldwin IT 1997 Herbivory and caterpillar regurgitants amplify the wound-induced increases in jasmonic acid but not nicotine in *Nicotiana sylvestris*. Planta 203:430–435

Powell W, Pennacchio F, Poppy GM, Tremblay E 1998 Strategies involved in the location of hosts by the parasitoid *Aphidius ervi* Haliday (Hymenoptera: Braconidae: Aphidiinae) Biol Control 11:104–112

Sabelis MW, Dicke M 1985 Long-range dispersal and searching behaviour. In: Helle W, Sabelis MW (eds) Spider mites: their biology, natural enemies and control, vol 1b. Elsevier Science, Amsterdam, (World Crop Pests) p 141–160

Shimoda T, Takabayashi J, Ashihara W, Takafuji A 1997 Response of predatory insect *Scolothrips takahashii* toward herbivore-induced plant volatiles under laboratory and field conditions. J Chem Ecol 23:2033–2048

Steinberg S, Dicke M, Vet LEM 1993 Relative importance of infochemicals from first and second trophic level in long-range host location by the larval parasitoid *Cotesia glomerata*. J Chem Ecol 19:47–59

Stout MJ, Workman J, Duffey SS 1994 Differential induction of tomato foliar proteins by arthropod herbivores. J Chem Ecol 20:2575–2594

Takabayashi J, Dicke M 1996 Plant–carnivore mutualism through herbivore-induced carnivore attractants. Trends Plant Sci 1:109–113

Takabayashi J, Dicke M, Posthumus MA 1991a Variation in composition of predator-attracting allelochemicals emitted by herbivore-infested plants: relative influence of plant and herbivore. Chemoecology 2:1–6

Takabayashi J, Dicke M, Posthumus MA 1991b Induction of indirect defence against spider-mites in uninfested lima bean leaves. Phytochemistry 30:1459–1462

Takabayashi J, Takahashi S, Dicke M, Posthumus MA 1995 Developmental stage of herbivore *Pseudaletia separata* affects production of herbivore-induced synomone by corn plants. J Chem Ecol 21:273–287

Turlings TCJ, Tumlinson JH 1992 Systemic release of chemical signals by herbivore-injured corn. Proc Natl Acad Sci USA 89:8399–8402

Turlings TCJ, Tumlinson JH, Lewis WJ 1990 Exploitation of herbivore-induced plant odors by host-seeking parasitic wasps. Science 250:1251–1253

Turlings TCJ, Tumlinson JH, Heath RR, Proveaux AT, Doolittle RE 1991 Isolation and identification of allelochemicals that attract the larval parasitoid, *Cotesia marginiventris* (Cresson), to the microhabitat of one of its hosts. J Chem Ecol 17:2235–2251

Turlings TCJ, Wäckers FL, Vet LEM, Lewis WJ, Tumlinson JH 1993 Learning of host-finding cues by hymenopterous parasitoids. In: Papaj DR, Lewis AC (eds) Insect learning: ecological and evolutionary pespectives. Chapman & Hall, New York, p 51–78

Turlings TCJ, Loughrin JH, McCall PJ, Röse USR, Lewis WJ, Tumlinson JH 1995 How caterpillar-damaged plants protect themselves by attracting parasitic wasps. Proc Natl Acad Sci USA 92:4169–4174

van Haren RJF, Steenhuis MM, Sabelis MW, De Ponti OMB 1987 Tomato stem trichomes and dispersal success of *Phytoseiulus persimilis* relative to its prey *Tetranychus urticae*. Exp Appl Acarol 3:115–121

Vet LEM 1999 Evolutionary aspects of plant–carnivore interactions. In: Insect–plant interactions and induced plant defence. Wiley, Chichester (Novartis Found Symp 223) p 3–20

Vet LEM, Dicke M 1992 Ecology of infochemical use by natural enemies in a tritrophic context. Annu Rev Entomol 37:141–172

Vet LEM, de Jong AG, Franchi E, Papaj DR 1998 The effect of complete versus incomplete information on odour discrimination in a parasitic wasp. Anim Behav 55:1271–1279

Visser JH, Ave DA 1978 General green leaf volatiles in the olfactory orientation of the Colorado beetle, *Leptinotarsa decemlineata*. Entomol Exp Appl 24:738–749

DISCUSSION

Poppy: Jim Tumlinson, I recall you saying that you found quite a lot of variation in herbivore saliva between individuals, and you were looking at bulking up the secretions from five individuals to overcome this. What sort of cross-reactivity is there between species? I've recently been discussing the salivary components of aphids with Freddy Tjallingii (Wageningen University, The Netherlands), to see whether the sophistication of saliva within aphids may explain why we see such aphid-specific plant responses. He was telling me that the monoclonal antibodies they have got for *Schizaphis graminum* don't cross-react with the saliva from a number of other aphid species (personal communication). There seems to be quite a lot of specificity within the saliva of aphids: what is the situation within the Lepidoptera?

Tumlinson: We haven't identified compounds from other Lepidopteran species apart from the beet armyworm, although we've taken a cursory look at several. We're working right now on *Manduca sexta*, and there appear to be similarities, but we have encountered some real difficulties working with *Manduca*. I think we can expect similarities among Lepidoptera, but when you get out of the Lepidoptera and start looking at grasshoppers or beetles, it could be quite different.

Poppy: It strikes me that with the Lepidoptera you've got this pair of jaws, with a range of enzymes, chomping through a plant. This could give a very different situation to that observed with an aphid, which is very careful about how it penetrates into the phloem. Therefore if it is the salivary components that are inducing various biosynthetic pathways within the plant, this would give you specificity with aphids that you might not get with the pair of jaws and battery of enzymes as observed in Lepidoptera.

Tumlinson: We know that parasitoids can tell the difference between a host Lepidopteran and a non-host Lepitopteran feeding on the same plant. We have also looked at the volatiles produced, and we see differences. Our problem is that we just haven't had time to identify the constituents of the spit, so we don't know what those differences are. We do know that there are about 8 or 10 compounds in the beet armyworm spit, and we know something about the activity of those compounds.

Dicke: But without knowing what's in the spit, you can't answer Guy Poppy's question about whether or not it is the combination of spit and type of mechanical damage that it is important here.

Tumlinson: You're absolutely right, there are those problems also. Without getting into chemically defined elicitors it's hard to answer that question.

Pickett: Just a point of clarification: Jim Tumlinson is talking about an organic compound which is not directly antigenic, whereas Guy Poppy is referring to enzymes which are, of course, antigenic.

Boland: Marcel Dicke, you argued that there's a negative correlation between direct and indirect effects. There is a problem with this, because both types of defence rely on signalling within the plant. Both types of defences are elicited, for example, by octadecanoid and hexadecanoid signalling. What therefore could be the basis for a different type of antagonistic pattern between the two defences?

Dicke: Even if they're signalling in the same way, then what you see is that in plants that are well known for a strong direct defence, there is not a very specific response in terms of the volatiles that are produced. For example, cabbage plants are well protected through glucosinolates, but when cabbage plants are damaged mechanically or by herbivores (or have regurgitate applied), the same kind of volatiles are released.

Boland: So the argument in this case is that there are metabolic pathways disconnected from the signalling pathway in such a way that they are not turned on.

Dicke: They are probably not disconnected actively by the individual plants, but instead over evolutionary history they have not been developed anymore, whereas they have been in plants like corn or bean that are less dependent on direct defences.

Of course, you could ask, do corn and bean not have any specific strongly defensive compounds? After all, phytoalexins are known from corn and bean. One of the problems is how we define a 'strong' defence: is it in terms of what compounds are made, or is it in terms of herbivores being affected?

Boland: Even cell wall reinforcement is a strong direct defence. This is also connected to the pathways related to damage.

Firn: I am a little disturbed by the fact that we have heard so little about dose–response curves so far. My naïve assumption is that not all compounds acting as attractants will be equally active. Let us assume that we have three compounds being produced in equal amounts by a plant, compounds which differ in their effectiveness as attractants as shown in Fig. 1 (*Firn*). The information content available to an insect will vary greatly depending on the distance between the insect and the plant. At point A on the dose–response curve, the insect will detect only one compound, at point B it will detect two and at point C all three, but at point D two of the compounds are producing a saturated response hence provide less information. Hence the testing of volatile mixtures surely must be done at a wide range of concentrations if we are seeking to understand how they influence insect behaviour. Presumably the way that the information content of the signal changes as the insect moves could be very important.

Dicke: In pheromone research it has been shown that there are puffs of sex pheromone that are being transported in little packages that don't change much

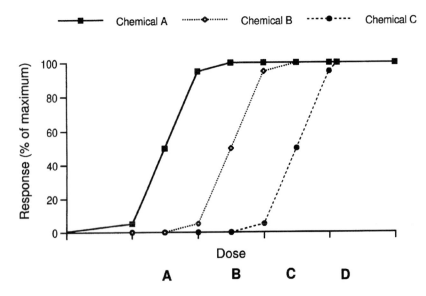

FIG. 1.   (*Firn*) Idealized dose–response curves for three attractants.

in composition. Only the frequency with which packages are received changes with distance from the source.

Firn: But that just changes where you are on the curve.

Pickett: There is only one graph. Low dose, high response at the single olfactory neuron level. The response does not extend over other compounds, although it can by EAG, but as Marcel Dicke said, that is because you have got different populations of olfactory neurons.

Wadhams: When an insect starts feeding on a plant, certainly with aphids, there is different temporal production of the various induced compounds. So there is the potential there, for example, with *Aphidius ervi*, for it to recognize host or non-host, and if it is a host aphid, how long that host has been on there and roughly what the population is going to be. We don't know whether that actually happens, but there is certainly that information content in the profile of volatiles.

Tumlinson: Over a short period, though, you should get essentially a consistent blend. But over 2–3 hours, the blend will change.

Wadhams: We see a great difference in temporal variability within a day. Over a longer period, such as 4–5 days, that is when we see some increased production of these compounds. There is potential for a lot of information being transmitted in the volatile blend.

Pickett: What about the very short burst?

Wadhams: That is the oxidative burst. Depending on the plant, this can be over and done in 5 minutes. But this is related purely to mechanical damage.

Karban: I would like to return to Marcel Dicke's expectation of a trade-off between direct and indirect defences. Even if we recognize a lot of specificity in responses, I don't know that we should expect trade-offs. In the system that I know best, the tomato work by Sean Duffey and his students, there is a lot of specificity in the responses of tomato plants even to different species of noctuid caterpillars. As a result of that specificity, perhaps it makes sense to assume little redundancy. Intuitively, it seems there may be some trade-off between induced and constitutive responses, but I haven't seen any evidence of that — in fact, I've seen evidence to the contrary. Since plants have evolved in environments with many herbivores and with the herbivores changing over time, it makes sense for plants to hold on to the ability to respond to all kinds of different insults. The expectation of parsimony in that plants get rid of the ones that are redundant is unfounded.

Dicke: But if you look at the example that you gave of tomato, where there are different responses to different noctuids in terms of which plant enzymes are switched on, that doesn't seem to be the case with volatile production. In tomatoes, mechanical damage and damage through spider mites results in the same blend (let's assume that the same machinery is behind it in both cases). This doesn't seem to be highly specific, whereas it is in the case of induced direct defence. This might indicate that the direct defence is much better developed in tomato.

Karban: It seems to me that there's a lot of imprecision. If we were to perfectly design plants to respond to herbivores, I'm not sure we would do it the way we see it happening in real plants. With or without the specificity of these responses, the expectation of trade-offs is one I no longer hold.

Lamb: I wonder whether there's a seasonal aspect to this. The parasitoid population develops after the herbivore population, so early in the season direct defences may have a non-redundant function relative to the indirect.

Vet: That could be very functional, if it really is the case.

Poppy: Aphid parasitoids will control aphids in the field if the synchrony is right. Part of the strategy that Wilf Powell and myself are trying to adapt at Rothamsted includes trying to manipulate the environment and/or the plant, or use aphid sex pheromones, to try to synchronize the populations and thus increase the impact of the parasitoids on the aphids.

Gershenzon: One good point that Marcel Dicke raised in his comparison of direct and indirect defences is that in species such as tomato, which have a high density of glandular trichomes (a direct defence), there may not be as much value in attracting carnivore species (indirect defence) if carnivores cannot operate freely due to the presence of trichomes. In this case there may be a clear trade-off between direct and indirect defences. Interestingly, cotton appears to have both direct and indirect defences without such a trade-off. There are strong direct defences, the gossypol-type terpenoid aldehydes in the sub-epidermal glands, which were shown to increase after damage (McAuslane & Alborn 1998). And, we've already heard from Jim Tumlinson about the indirect defences in this species. Maybe it's really a question of the mode of storage of direct defences. If they are deposited on the surface and somehow interfere with the ability of carnivores and parasitoids to operate, then there is a trade-off. But, if they're internally stored, it may be possible to exploit both direct and indirect modes of defence at the same time.

Dicke: There is an example of these trade-offs that has nothing to do with volatiles, which is the work by Vrieling et al (1991) on *Senecio jacobaeae*, *Tyria* caterpillars, ants and aphids. Plant genotypes that have low amounts of pyrrolizidine alkaloids (direct defence) have a lot of aphids and consequently a lot of ants that collect honeydew. As a result those plants harbour few *Tyria jacobaeae* caterpillars, which are eliminated by the ants (indirect defence). These plants thus have a weak direct defence and a strong indirect defence. The genotypes that have a lot of pyrrolizidine alkaloids have no or few aphids (aphids avoid the alkaloids) and consequently no or few ants and thus a weak indirect defence against *Tyria* caterpillars; they have a strong direct defence but not against *Tyria* caterpillars because these specialist herbivores can cope with the alkaloids. These genotypes are completely munched up by the caterpillars when their densities are high, but they survive in years when the caterpillar populations are low. This is a clear example of a trade-off between direct and indirect defence.

References

McAuslane HJ, Alborn HT 1998 Systemic induction of allelochemicals in glanded and glandless isogenic cotton by *Spodoptera exigua* feeding. J Chem Ecol 24:399–416

Vrieling K, Smit W, van der Meijden E 1991 Tritrophic interactions between aphids (*Aphis jacobaeae* Schrank), ant species, *Tyria jacobaeae* L. and *Senecio jacobaea* L. lead to maintenance of genetic variation in pyrrolizidine alkaloid concentration. Oecologia 86:177–182

Aphids, predators and parasitoids

Lester J. Wadhams, Michael A. Birkett, Wilf Powell* and Christine M. Woodcock

*Biological and Ecological Chemistry Department and *Entomology and Nematology Department, IACR-Rothamsted, Harpenden, Hertfordshire AL5 2JQ, UK*

Abstract. A number of studies have demonstrated the role of herbivore-induced release of plant volatiles in mediating foraging behaviour of aphid parasitoids, particularly with the parasitoid *Aphidius ervi*, its aphid host *Acyrthosiphon pisum* and the aphid food plant *Vicia faba*. These studies have shown that feeding by the aphid alters the composition of volatiles released by the plant and that these compounds act as synomones for the foraging parasitoid. Of particular interest is the species-specificity of the herbivore-induced synomones associated with different aphids feeding on *V. faba*. Aphids employ various pheromones that mediate behaviour, particularly mating and alarm responses. These pheromones play important roles in reproduction and defence against predation and parasitism. Many species of aphids reproduce sexually on their primary hosts during the autumn and the sexual females produce a pheromone that attracts males. The sex pheromones for a number of aphid species have been identified and laboratory and field studies have shown that synthetic material can act as a kairomone in attracting predators and parasitoids. The aphid alarm pheromone is released from the cornicles of aphids when they are attacked by predators or parasitoids. The activity of the main alarm pheromone component, (E)-β-farnesene, is inhibited by the related sesquiterpene hydrocarbon β-caryophyllene, which is reported to attract the lacewing *Chrysoperla carnea*. In addition, electrophysiological studies have shown that the seven-spot ladybird, *Coccinella septempunctata*, possesses specific olfactory receptors for (E)-β-farnesene and β-caryophyllene. Laboratory studies show these compounds to have behavioural activity with *C. septempunctata*, suggesting that they may be involved in prey location.

1999 Insect–plant interactions and induced plant defence. Wiley, Chichester (Novartis Foundation Symposium 223) p 60–73

A wide range of beneficial organisms, including insect predators and parasitic wasps, help to regulate populations of aphid pests. It has been shown that cues mediating foraging behaviour of female aphid parasitoids can be derived either from the host insect or from the food plant of the host, or may arise from interactions between the two. However, very little is yet known about the semiochemicals involved in prey location by aphidophagous insects such as ladybird beetles and lacewings.

Plant volatiles as foraging cues

Recent studies, using a model tritrophic system comprising *Aphidius ervi* (parasitoid), *Acyrthosiphon pisum* (host aphid), and broad bean *Vicia faba* (aphid food plant), have demonstrated the crucial role of aphid-induced release of plant volatiles in mediating the foraging behaviour of aphid parasitoids (Guerrieri et al 1993, Du et al 1996, Powell et al 1998). In wind tunnel bioassays, naïve *A. ervi* females responded poorly to undamaged bean plants and to aphids isolated from the plant. However, they showed strong oriented flight behaviour towards aphid-infested plants and also towards aphid-damaged plants, even after the aphids had been removed and the leaves washed with water to remove exuviae and honeydew. In addition, plants with aphid infestations confined to the basal leaf, which was subsequently removed before the bioassay, remained attractive to female parasitoids for up to 24 h, even though aphids had never been in contact with the remaining parts of the plant. These results strongly suggest that feeding by the aphid alters the composition and/or ratios of volatiles released by the plant and that herbivore-induced plant volatiles act as synomones for the foraging parasitoid. However, many aphid parasitoids have restricted host ranges and the same plant species may be the food plant for both host and non-host aphids. Thus, *V. faba* plants are commonly infested by the black bean aphid, *Aphis fabae*, which is not attacked by *A. ervi*, and in wind tunnel tests the parasitoid did not distinguish between uninfested and *A. fabae* infested plants. In contrast, female parasitoids did land preferentially on *A. pisum* infested plants when given a choice between these and plants infested with *A. fabae*, indicating that they could, at a distance, distinguish between plants infested with hosts and those infested with non-host aphids (Du et al 1996, Powell et al 1998). This suggests that the aphid-induced plant volatiles in this model system are herbivore species-specific, with *A. fabae* not stimulating production of the same blend of plant volatiles as is induced by *A. pisum* and which is used as a host location cue by *A. ervi* females. This conclusion is supported by chemical evidence. Volatile extracts entrained from plants infested with *A. pisum* elicited significantly stronger responses from female *A. ervi*, in both olfactometer and wind tunnel bioassays, than extracts from uninfested or *A. fabae* infested plants (Du et al 1998, Powell et al 1998).

Gas chromatography coupled with electroantennogram recordings (GC-EAG) from *A. ervi* females revealed a number of active components associated with the extract of *V. faba* infested with *A. pisum*. Although many of these were identified as typical plant volatiles, they have been implicated as behavioural cues in other trophic systems, e.g. based on corn, lima bean or cucumber (Dicke et al 1990a,b, Turlings et al 1991), and there was an increase in their levels in the extract from *A. pisum* infested broad bean plants, compared to those in extracts from uninfested plants. However, one compound, 6-methyl-5-hepten-2-one, which had not previously been reported as resulting from herbivore feeding damage, did show a difference between extracts. As demonstrated by GC analysis, its release from *V. faba* was induced by *A. pisum* but not by *A. fabae* feeding damage. When compounds identified from the *A. pisum/V. faba* complex were tested in wind tunnel bioassays, synthetic 6-methyl-5-hepten-2-one was the most attractive to naïve *A. ervi* females, eliciting significantly more landing responses than any other compound (Du et al 1998). Thus, this compound is potentially one of the volatile components that allows *A. ervi* to distinguish between host and non-host aphid-infested plants, but this requires verification in further work.

For parasitoids such as *A. ervi*, which has a limited host range, the ability to distinguish, whilst in flight, between plants infested with appropriate or inappropriate hosts would be advantageous. *A. ervi* is oligophagous in its host range, attacking aphids normally associated with leguminous plants such as *A. pisum*, but also attacking several cereal aphids (Starý 1973, Powell 1982). Indeed, in parts of South America, it has become an important control agent for cereal aphids following its introduction and release some years ago (Starý 1993). Although legumes and cereals appear to be very different host plants, chemical studies have shown that 6-methyl-5-hepten-2-one is also present in the volatiles produced by aphid-damaged cereal plants (Quiroz et al 1997), suggesting that this compound may have a wider role in mediating host location by *A. ervi* than has so far been demonstrated.

Parasitoid responses to semiochemicals are complex processes and many parasitoids have developed learning abilities, based on foraging experience, which allow them to modify their responses to foraging cues (Turlings et al 1993, Vet et al 1995). The ability to learn as a result of previous experience endows parasitoids with a degree of behavioural plasticity which enables them to adapt to changing foraging opportunities, such as alterations in the relative abundances of alternative host species. Allowing *A. ervi* females to forage on the *A. pisum*/*V. faba* host/plant complex before testing in behavioural bioassays significantly increased their responses to this complex (Du et al 1997, Guerrieri et al 1997), and also to the individual compounds identified from volatile extracts as being associated with *A. pisum* feeding damage to *V. faba* (Du et al 1998). The exception to this was (E)-β-farnesene (EBF), which is also produced by many economically important aphid species, including *A. pisum*, as the major component of an alarm pheromone released in response to attack by natural enemies and causing dispersal of nearby conspecific aphids (Pickett et al 1992). Host pheromones are frequently used by parasitoids as host location cues (functioning as kairomones) (Powell 1999), and whilst wind tunnel studies have shown that naïve female *A. ervi* respond to EBF, this response was not enhanced by previous experience with the plant/host complex (Du et al 1998). It has been proposed that parasitoid learning processes, and the resultant behavioural plasticity, are most likely to involve responses to less reliable foraging cues, such as plant-derived synomones, when these are associated with host-derived kairomones eliciting strong innate (fixed) responses (Vet & Dicke 1992, Turlings et al 1993). Although EBF is released by plants, its role as an aphid alarm pheromone may mean that, in the case of aphid parasitoids, it functions principally as a host-derived behavioural cue. Thus, responses to EBF may be more fixed and therefore less likely to change as a result of learning processes.

Aphid pheromones as foraging cues

Ladybirds, particularly the seven-spot ladybird, *Coccinella septempunctata*, are important aphid predators and in behavioural studies, adults were strongly attracted towards aphid odour sources in an olfactometer (Sengonca & Liu 1994). Electrophysiological studies on *C. septempunctata* antennae showed the presence of specific olfactory receptors responding to EBF, a compound emitted by both plants and aphids, and the biosynthetically related β-caryophyllene (Birkett et al 1999). Dose–reponses of the two cell types showed similar sensitivities. These receptors were frequently found to occur in pairs, with typically the large amplitude cell responding to EBF, suggesting their co-location on the same olfactory sensillum. In behavioural studies, EBF was strongly attractive to *C. septempunctata* adults; however, addition of increasing proportions of β-caryophyllene resulted in a decrease in the response. This is analogous to the situation in aphids, where the response to EBF, when released as an alarm pheromone by aphids, is inhibited by the presence of β-caryophyllene, presumably as a mechanism such that alarm behaviour is not induced merely by the release of EBF from a plant (Dawson et al 1984). Perception of these two compounds by aphids is again mediated by specialized olfactory cells on the antenna.

Lacewing larvae are considered to be one of the most effective general entomophagous predators and some species prefer aphids to other prey when offered a choice. It has been suggested that the aphid alarm pheromone may attract adult lacewings and, indeed, electrophysiological recordings from the antennae of *Chrysoperla carnea* have demonstrated their ability to detect EBF (Fig. 1). Although no behavioural studies have yet been conducted with EBF, β-caryophyllene has been reported to attract *C. carnea* (Flint et al 1979). In contrast, studies with *Chrysopa cognata*, one of the commonest lacewings in Korea and which also shows a preference for aphids, showed no electrophysiological or behavioural response to the aphid alarm pheromone (Boo et al 1998).

Many species of aphids reproduce sexually on their primary hosts during the autumn and the sexual females (oviparae) produce a pheromone that attracts males (Pettersson 1970, 1971). The sex pheromones for a number of aphid species have been identified and shown to comprise one or both of (4aS,7S,7aR)-nepetalactone and the corresponding (1R,4aS,7S,7aR)-nepetalactol (Pickett et al 1992). Although these pheromones are released only over a limited period in the

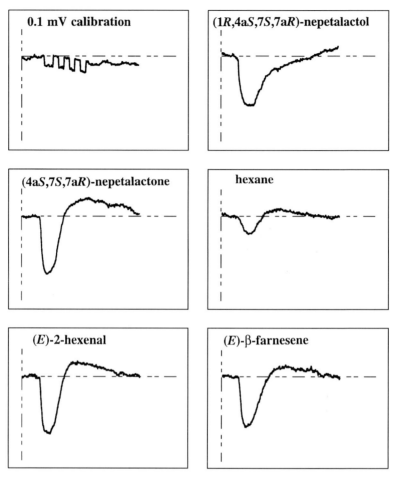

FIG. 1.   EAG responses of adult green lacewing *Chrysoperla carnea* (compounds at 10^{-5} g in hexane).

autumn, electrophysiological studies have shown that a wide range of aphidophagous predators, including lacewings and ladybirds, are able to detect these compounds (e.g. Fig. 1), suggesting that they play a role in the chemical ecology of these predators. Indeed, behavioural studies with *C. cognata*, both in the laboratory and in the field, have shown strong attraction of this lacewing to the pheromone components, particularly the nepetalactone (Boo et al 1998). It is somewhat unexpected that a generalist aphid predator should utilize, as a kairomone, such a specific prey pheromone. It has been suggested that generalist predators might respond to more common biochemicals such as amino acids,

TABLE 1   EAG responses of female aphid parasitoids to (4aS,7S,7aR)-nepetalactone at 10^{-6} g in hexane (data otherwise unpublished)

	Response (mV)[a]		
Species	Treatment	Control[b]	P[c]
Praon volucre	1.23 ± 0.03	0.15 ± 0.05	<0.001
Aphidius ervi	1.60 ± 0.20	0.07 ± 0.01	<0.001
A. rhopalosiphi	0.48 ± 0.01	0.05 ± 0.01	<0.001
A. matricariae	0.47 ± 0.06	0.06 ± 0.01	<0.001
A. urticae	0.41 ± 0.03	0.04 ± 0.005	<0.001
Diaretiella rapae	0.18 ± 0.02	0.01 ± 0.006	<0.001

[a]Means of 5 preparations $\pm$ SEM.
[b]Control=hexane.
[c]Significance of difference calculated using Student's t-test.

whereas more specialized predators may respond to semiochemicals more closely associated with their hosts or the host/plant complex (Greany & Hagen 1981). However, the fact that a predator is relatively polyphagous does not preclude its use of specific semiochemical cues for prey finding and, indeed, there are reports that generalist feeders are also attracted to odours from their prey (Greany & Hagen 1981).

Trap catches of female *Praon* spp. parasitoids in the field were particularly associated with the nepetalactone component of aphid sex pheromones (Hardie et al 1991, 1994, Powell et al 1993). Indeed, this compound elicits significant EAG responses with females of all the parasitoid species so far investigated (Table 1). Although *A. ervi* females, which have a more limited host range than *Praon* spp., have not been caught in the field in traps releasing aphid sex pheromone components, they will respond to filter paper targets treated with synthetic pheromones in the wind tunnel. The strongest orientated flight and landing responses occurred with a 1:1 mixture of nepetalactone:nepetalactol. In the field, using potted plants with or without aphid sex pheromone lures, significantly more *A. pisum* were parasitized by *A. ervi* and *Aphidius eadyi* on the treated plants than on the controls (Powell et al 1998).

Conclusions

Aphid predators and parasitoids have considerable potential as biological control agents in the development of novel crop protection strategies, but their efficiency is dependent on their presence at the right time and in the right place. Our understanding of the mechanisms mediating host location by predators and parasitoids, together with the identification of the semiochemical cues regulating this behaviour, is providing exciting opportunities for the manipulation of natural populations of these beneficial organisms in the field.

Acknowledgements

IACR-Rothamsted receives grant-aided support from the Biotechnology and Biological Sciences Research Council, UK. This work was in part supported by the Ministry of Agriculture, Fisheries and Food, UK.

References

Birkett MA, Pickett JA, Wadhams LJ, Woodcock CM, Al Abassi S, Pettersson J 1999 Electrophysiological and behavioural responses of the seven-spot ladybird, *Coccinella septempunctata* L. (Coleoptera, Coccinellidae), to *(E)-β*-farnesene, the aphid alarm pheromone, and β-caryophyllene. In prep

Boo KS, Chung IB, Han KS, Pickett JA, Wadhams LJ 1998 Response of the lacewing *Chrysopa cognata* to pheromones of its aphid prey. J Chem Ecol 24:631–643

Dawson GW, Griffiths DC, Pickett JA, Smith MC, Woodcock CM 1984 Natural inhibition of the aphid alarm pheromone. Entomol Exp Appl 36:197–199

Dicke M, van Beek TA, Posthumus MA, Ben Dom, N, Van Bokhoven H, De Groot AE 1990a Isolation and identification of volatile kairomone that affects acarine predator prey interactions. Involvement of host plant in its production. J Chem Ecol 16:381–396

Dicke M, Sabelis MW, Takabayashi J, Bruin J, Posthumus MA 1990b Plant strategies of manipulating predator–prey interactions through allelochemicals: prospects for application in pest control. J Chem Ecol 16:3091–3118

Du Y-J, Poppy GM, Powell W 1996 Relative importance of semiochemicals from the first and second trophic level in host foraging behavior of *Aphidius ervi*. J Chem Ecol 22:1591–1605

Du Y-J, Poppy GM, Powell W, Wadhams LJ 1997 Chemically mediated associative learning in the host foraging behavior of the aphid parasitoid *Aphidius ervi* (Hymenoptera: Braconidae). J Insect Behav 10:509–522

Du Y-J, Poppy GM, Powell W, Pickett JA, Wadhams LJ, Woodcock CM 1998 Identification of semiochemicals released during aphid feeding that attract parasitoid *Aphidius ervi*. J Chem Ecol 24:1355–1368

Flint HM, Salter SS, Walters S 1979 Caryophyllene: an attractant for the green lacewing. Environ Entomol 8:1123–1125

Greany PD, Hagen KS 1981 Prey selection. In: Nordlund DA, Jones RJ, Lewis WJ (eds) Semiochemicals: their role in pest control. Wiley, Chichester, p 51–77

Guerrieri E, Pennacchio F, Tremblay E 1993 Flight behaviour of the aphid parasitoid *Aphidius ervi* (Hymenoptera: Braconidae) in response to plant and host volatiles. Euro J Entomol 90:415–421

Guerrieri E, Pennacchio F, Tremblay E 1997 Effect of adult experience on in-flight orientation to plant and plant–host complex volatiles in *Aphidius ervi* Haliday (Hymenoptera, Braconidae). Biol Control 10:159–165

Hardie J, Nottingham SF, Powell W, Wadhams LJ 1991 Synthetic aphid sex pheromone lures female parasitoids. Entomol Exp Appl 61:97–99

Hardie J, Hick AJ, Höller C et al 1994 The responses of *Praon* spp. parasitoids to aphid sex pheromone components in the field. Entomol Exp Appl 71:95–99

Pettersson J 1970 An aphid sex attractant. I. Biological studies. Entomol Scand 1:63–73

Pettersson J 1971 An aphid sex attractant. II. Histological, ethological and comparative studies. Entomol Scand 2:81–93

Pickett JA, Wadhams LJ, Woodcock CM, Hardie J 1992 The chemical ecology of aphids. Annu Rev Entomol 37:67–90

Powell W 1982 The identification of hymenopterous parasitoids attacking cereal aphids in Britain. Syst Entomol 7:465–473

Powell W 1999 Insect pheromones as foraging cues for parasitoids. In: Hardie RJ, Minks AK (eds) Pheromones of non-Lepidopteran insects associated with agricultural crops. CAB International, Wallingford, in press

Powell W, Hardie J, Hick AJ et al 1993 Responses of the parasitoid *Praon volucre* (Hymenoptera: Braconidae) to aphid sex pheromone lures in cereal fields in autumn: implications for parasitoid manipulation. Eur J Entomol 90:435–438

Powell W, Pennacchio F, Poppy GM, Tremblay E 1998 Strategies involved in the location of hosts by the parasitoid *Aphidius ervi* Haliday (Hymenoptera: Braconidae: Aphidiinae). Biol Control 11:104–112

Quiroz A, Pettersson J, Pickett JA, Wadhams LJ, Niemeyer HM 1997 Semiochemicals mediating spacing behavior of bird cherry-oat aphid, *Rhopalosiphum padi*, feeding on cereals. J Chem Ecol 23:2599–2607

Sengonca C, Liu B 1994 Responses of the different instar predator, *Coccinella septempunctata* L. (Coleoptera, Coccinellidae), to the kairomones produced by the prey and non-prey insects as well as the predator itself. J Plant Dis Prot 101:173–177

Starý P 1973 A review of *Aphidius*-species (Hymenoptera: Aphidiidae) of Europe. Annot Zool Bot 84:1–85

Starý P 1993 The fate of released parasitoids (Hymenoptera: Braconidae, Aphidiinae) for biological control of aphids in Chile. Bull Entomol Res 83:633–639

Turlings TCJ, Tumlinson JH, Heath RR, Proveaux AT, Doolittle RE 1991 Isolation and identification of allelochemicals that attract the larval parasitoid, *Cotesia marginiventris* (Cresson), to the microhabitat of one of its hosts. J Chem Ecol 17:2235–2251

Turlings TCJ, Wäckers FL, Vet LEM, Lewis WJ, Tumlinson JH 1993 Learning of host-finding cues by Hymenopterous parasitoids. In: Papaj DR, Lewis AC (eds) Insect learning: ecological and evolutionary aspects. Chapman & Hall, New York, p 51–78

Vet LEM, Dicke M 1992 Ecology of infochemical use by natural enemies in a tritrophic context. Annu Rev Entomol 37:141–172

Vet LEM, Lewis WJ, Cardé RT 1995 Parasitoid foraging and learning. In: Cardé RT, Bell WJ (eds) Chemical ecology of insects 2. Chapman & Hall, New York, p 65–101

DISCUSSION

Vet: If I understood correctly, *Praon* (a generalist parasitoid) is caught in the aphid pheromone trap, but *Aphidius ervi* is not. Is this because *A. ervi*, as a specialist, needs other cues and not just the aphid pheromone?

Wadhams: These aphid sex pheromones are released in the autumn and represent the last chance for the parasitoid to oviposit an egg before the winter, so there is a high drive for them to respond. *Praon* can oviposit in a wide range of aphids, but

A. ervi can't — it appears to use the signal rather as an arrestant. There are other possibilities: maybe they both have different thresholds.

Vet: Do you only find this effect in the autumn?

Wadhams: No, the field trials were done in the summer as well.

Poppy: Praon species are generalist aphid parasitoids and thus need to find an aphid, so the pheromone is a reliable cue as the majority of aphid species studied have the same two pheromone components (nepetalactone and nepetalactol), in differing ratios. However, *A. ervi* can't totally rely on the aphid sex pheromone, unless it can determine the exact ratios, otherwise it could be attracted to the wrong aphid species and thus waste foraging time.

Wadhams: But these parasitoids are still responding in the summer to what is essentially a biologically irrelevant cue, since the sex pheromone is only produced by the aphids in the autumn.

Poppy: In New Zealand there are populations of aphids which never go sexual whereas populations of the same species in England do. The parasitoids which were taken out there 20 or so years ago were from Europe. It would therefore be very interesting to look at that population which has not encountered aphid sex pheromones at all over the last 20 years.

Eastop: There is a problem there that in the south of South Island, some aphids overwinter sexually.

Turlings: I have a question about caryophyllene and its effect. Is there any biological reason why caryophyllene would reduce the responsiveness to EBF?

Wadhams: The aphid has to be able to distinguish between aphid-derived EBF and plant-derived EBF. This is achieved by monitoring the ratios of EBF and caryophyllene, for which the aphid has separate olfactory systems. High proportions of caryophyllene, as found in plants, reduce the aphid alarm pheromone response. We have to remember that ladybirds are polyphagous, but the fact that they also possess specific olfactory cells for EBF and caryophyllene does suggest that, in a similar manner to aphids, ladybirds are trying to determine whether the EBF that they're perceiving on their antennae is derived from a plant source or from an aphid prey source.

Turlings: It is interesting also that with respect to the induced volatiles that we observe from corn and other grasses, there we do see both EBF and caryophyllene, and some corn varieties do not produce caryophyllene and others do. It would be interesting to compare these and see their effects on the parasitoids and predators.

Firn: The aphid presumably gets a lot of information from the temporal changes in a signal. An alarm pheromone will suddenly rise in concentration. Therefore, if an aphid can measure the rate of change of a signal from individual volatiles it will be able to discriminate between plant-derived and insect-derived chemicals.

Wadhams: The only problem with that is that we don't know what's happening at the boundary layer, because the aphids are below the boundary layer most of the time. There may be differences there in the diffusion of molecules.

Pickett: We did a lot of bioassay work in this area. The inhibitory effect of the caryophyllene can be overcome by a sudden increase in EBF concentration. What you say is true, but there is a recognition issue also, because if the situation is reversed, e.g. by letting a predatory ladybird loose on the aphids together with a background of caryophyllene, then the aphids do not disperse as effectively. There is recognition of the compound itself, but the aphids also recognize the rapid increase in concentration of EBF. There are other sesquiterpenes that also interfere, but caryophyllene seems to satisfy the ecological issue here.

Firn: It is clear that insects can derive much useful information from sensing temporal changes in volatile signal strength. Hence we should be cautious in using time-averaged analytical approaches when measuring volatiles.

Poppy: I have a question about this pairing of peripheral cells. At the recent Insect–Plant Relationships conference in Oxford (SIP 10, July 1998), Gilles Laurent (Caltech) gave a plenary talk on his work investigating the central integration of olfactory responses in insects. He showed that there was so much processing of signals within the brain, for example from interneurons back into the mushroom bodies, and the recognition of time codes (Laurent 1997), that you could get recognition of different olfactory signals in the brain, even when the same receptor is firing and no peripheral differences are observed. The number of paired cells being picked up is relatively sparse. This might not be the only mechanism for actually doing this: it may be occurring nearly all the time, but we don't know enough about the processing in the brain to pick it up.

Städler: That is true. There is the possibility that there is even an interaction within the sensillum between the two cell bodies. There are now quite a few examples in pheromone perception that the message is decoded, and is decoded in terms of relations between cell firing.

Wadhams: I agree. We are just looking in simplistic terms at the moment: I'm sure it's much more sophisticated than that. Don't forget that the placoid sensilla in Hymenoptera probably offer the biggest chance of this cell pairing, because you have got 80 or 90 olfactory neurons on one sensillum at any one time.

Firn: Could one breed populations of aphids and select the mutants which are deficient in their ability to sense chemical signals?

Wadhams: I'm sure you could.

Pickett: It is certainly possible to breed pesticide-resistant aphids.

Wadhams: That's at the behavioural level: at the physiological level we don't see a difference.

Städler: It wouldn't be easy to do recordings and then tell this aphid to go and reproduce. It is obviously much easier to select by using insecticide.

Wadhams: No, you can actually do that, because frequently when we're doing these single cell recordings the aphid is actually reproducing at the time.

Firn: A naïve more general question: if you go out into a field or forest and trap the volatiles, what kind of mixes did you find? How complex are they compared with your laboratory situations?

Wadhams: That's something we're trying to develop. We are becoming concerned that the plants that we're dealing with in the lab, grown in the greenhouse, don't represent what's happening in nature. This approach is now becoming feasible, but the problem is that there is a vast amount of contamination out there.

Firn: But that's the real world.

Wadhams: Yes, that's the real world, and that's the world that they function in, but it's very difficult for us to pick out from that complex mixture those compounds that are relevant to the insect. To do this you have to use the coupled electrophysiology system and this means that you have to be able to trap reproducibly what's happening in the environment.

Poppy: Whilst the real world is far more complex than the laboratory situation, when we do the behavioural tests we give the parasitoid only a single opportunity to respond, whereas in the real world the parasitoid may have hundreds of attempts. One of the reasons we try to simplify things from the behavioural perspective is to give clarity and to allow us to know in which direction to move forward. If we tried to publish that we put a plant at one end of a wind tunnel and came back four days later to see whether an insect had landed on it, the journal's referees wouldn't be impressed!

Turlings: I guess Richard Firn's question is related to whether or not what we see in the laboratory can be translated into field, and whether the other odours present may be masking the odours emitted by the plants to attract the parasitoids. If we take the examples from pheromone research, a lot less is emitted and the insects are very capable of finding each other using similar compounds in many cases. On this evidence, I'd say that the parasitoids would also be able to pick up the signals.

Firn: This suggests once again that insects must have very high powers of discrimination in terms of volatile signal detection.

Pickett: The practical development of field entrainment equipment is something we see as an important part of our field programme. We need a better idea of what's going on in standing field crops. These aphid parasitoids have wonderful molecular recognition equipment which operates in a tremendously competitive environment. It is very difficult to make an analogue of a pheromone or of a plant-derived semiochemical that gives a significant response with the single neuron that is tuned to the particular natural product. Usually several orders of magnitude more of the analogue is required to give a comparable effect. It is this type of molecular recognition that underpins all these types of interactions in chemical ecology.

Firn: Presumably there is some behavioural variability in the parasitoid population.

Poppy: As Marcel Dicke alluded to, phenotypic plasticity is vast. The genotypic variability is, too, but it is so far untapped, although our lab is becoming interested in molecular markers to determine behavioural traits.

Dicke: We know that the phenotypic components can be quite variable depending on what each individual has experienced in the previous hours or days.

Pickett: For some organisms it must be evolutionarily advantageous to have behavioural variability. For example, for a terrestrial mollusc with no shell, then by all (i.e. 100%) coming out to forage on detection of a plant cue then, if coincident with high bird predation, there would be a dramatic reduction in population.

Firn: If there is a fair degree of behavioural variability in a population, this might frustrate our attempts to manipulate plant–insect interactions. Essentially, survivors with a behaviour that frustrates our control measures will soon thrive.

Poppy: One of the reasons we're keen on using the aphid sex pheromone for manipulating aphid parasitoids in the field is that there we have an innate response to this pheromone which doesn't seem to be plastic.

Wadhams: But also this is only one component of an overall strategy. We're trying to ensure that none of the components is that effective in its own right, so that at no stage are we imposing high selection pressures on populations, particularly in the case of aphids where reproduction rates are phenomenally high. We want the *overall* strategy to maintain pest populations below economic threshold levels.

Pickett: In all of our practical programmes we do not envisage using just one semiochemical type: we try to have a number of components integrated also with a direct population-reducing component such as a fungal pathogen. Hence this push–pull idea that Jim Miller and others originally advanced (Miller & Cowles 1990). By this means we can put together a series of say 30% effects on population levels at the same time, just using what could occur naturally. This should ensure that there is no great selection pressure on any one semiochemical type. Our programme in Kenya will eventually expand over Africa, and this will be an interesting area in which to see whether resistance arises to such an integrated semiochemical-based strategy.

Schultz: For those of you who work with aphids, do you ever see what a plant pathologist would refer to as a 'hypersensitive response', particularly at the point of feeding?

Wadhams: Not that I am aware of.

Schultz: Connected with that, I have a comment on aphids. I would not be so bold as to suggest that what a plant does in response to aphids is directly related to the amount of damage that they do. They are in a position to manipulate responses, and the amount of tissue destroyed is probably no measure of what the plant does in response to the aphids. As we find out more about what aphids are doing, I think we will find that they understand plant signalling better than we do and are turning it on or off when they want to.

Dicke: It's clear that you don't need any damage to induce a response. With regard to hypersensitive responses, there have been recordings of these to *Pieris* eggs on *Brassica* species back in the 1980s (Shapiro & DeVay 1987).

Schultz: And of course, once you have something that resembles a hypersensitive response and an oxidative burst, you can see development of systemic responses that are dramatic without terribly much damage.

Firn: Following aphid attack, is virus infection or sucrose loss the biggest problem for the plants?

Poppy: It depends who you talk to. Pathologists at Rothamsted say that the only damaging thing about an aphid is that it transmits a virus! Getting back to the aphids, that is why we are hoping to start collaborating with Freddie Tjallingii to look at this aspect. We want to interrupt the aphid's feeding at various stages, for instance just after the initial penetration, during which watery saliva will have entered the phloem. By doing this we would hope to see when the aphids actually cause the plant to signal and attract the parasites.

Schultz: This might be an interesting place to employ some molecular probes stolen from plant pathology. All aphids contain symbiotic bacteria, and bacterial products are excellent elicitors of some kinds of responses. I don't think anyone is approaching this interface from this point of view.

Mitchell-Olds: Pieterse et al (1998) have shown that a nematode resistance gene in tomato also causes resistance to aphids. However, the mechanism is unknown.

Pickett: Yes, I have heard this with some surprise: many of our transgenic plants have more aphids on them then the controls!

Schultz: One thing to do would be to clean out the aphid and see whether what they do with the plant is maintained.

Pickett: The problem would be that without the symbiotic bacteria the aphids would die. But to answer your question in more detail, aphids cause damage at various levels. Aphids can destroy field beans and Brussels sprouts completely by direct feeding. In cereals in the autumn, pesticides are sprayed against aphids transmitting barley yellow dwarf virus, whereas in the summer there may be treatment against aphids which cause cosmetic damage or possibly health problems by encouraging development of fungi.

Firn: Does virus infection occur in the first few minutes of an aphid feeding?

Pickett: Not necessarily. There are formally three kinds of transmission: persistent, semi-persistent and non-persistent. For example, potato virus Y is non-persistently transmitted and this occurs very quickly, whereas barley yellow dwarf virus takes many hours of feeding to be transmitted by the persistent mode.

Firn: The reason I thought about the speed of viral infection is that having some parasitoid coming along to get the aphid is not going to be much value to the plant if the plant has already suffered its loss. Are there any differences in terms of volatile production in the plants subjected to different types of attack?

Pickett: That is interesting, because we have a guild of cereal aphids, some of which are transmitting viruses and some of which are not; it would be possible to explore that.

Poppy: There's the work of Ian Williams, who did a PhD with Alan Dewar at IACR Brooms Barn, showing that when both *Myzis persicae* and *Aphis fabae* were feeding on sugarbeet, *M. persicae* aphids were altering the secondary metabolism of the plant in such a way that other *M. persicae* did better than *A. fabae*, and vice versa (Williams et al 1999). The aphids are altering the plant secondary metabolism for their own benefit, and this is something the parasitoids then exploit. Over evolutionary time the aphids are not going to stop doing this because it is something they are doing to benefit themselves.

References

Laurent G 1997 Olfactory processing: maps, time and codes. Curr Opin Neurobiol 7:547–553

Miller JR, Cowles RS 1990 Stimulo-deterrent diversion: a concept and its possible application to onion maggot control. J Chem Ecol 16:3197–3212

Pieterse C, van Wees S, van Pelt J et al 1998 A novel signaling pathway controlling induced systemic resistance in *Arabidopsis*. Plant Cell 10:1571–1580

Shapiro AM, DeVay JE 1987 Hypersensitivity reaction of *Brassica nigra* L. (Cruciferae) kills eggs of *Pieris* butterflies (Lepidoptera: Pieridae). Oecologia 71:631–632

Williams IS, Dewar AM, Dixon AFG 1999 Positive feedback in aphid–plant interactions. Entomol Exp Appl, in press

Functional interactions in the use of direct and indirect defences in native *Nicotiana* plants

Ian T. Baldwin

Max-Planck-Institut für Chemische Ökologie, Tatzendpromenade 1a, D-07745 Jena, Germany

Abstract. *Nicotiana attenuata* has both direct (induced nicotine production) and indirect (induced release of mono- and sesquiterpenes) defences induced by herbivore attack; both are activated by the jasmonate cascade, albeit in different tissues (roots and shoots, respectively). The fact that both types of defences are induced suggests that their benefits are conditional. Indeed, jasmonate treatment of roots to induce nicotine production increases plant fitness correlates (lifetime viable seed production) when plants are grown in environments with herbivores, but decreases fitness when they are not. Because inducing nicotine production can make 6% of a plant's nitrogen budget unavailable for seed production, it can exact a resource-based cost. Volatile production is likely to be less costly but could make plants more 'apparent' to herbivores and thereby exact an ecological cost. Direct defences could also have ecological costs if they are sequestered by specialist herbivores and used against their enemies. Herbivory by the nicotine-tolerant herbivore *Manduca sexta* dramatically amplifies the increase in jasmonates and the quantity of volatiles released, but decreases the nicotine response in comparison to mechanical simulations of the wounding that larval feeding causes. The apparent switching from nicotine production to the release of volatiles may reflect incompatibilites in the use of direct and indirect defences with specialist herbivores.

1999 Insect–plant interactions and induced plant defence. Wiley, Chichester (Novartis Foundation Symposium 223) p 74–94

Herbivores have been a powerful selective force for plants, and an explicit research objective of plant biologists for more than a century has been to identify those plant traits which result from this interaction (Stahl 1888). The response of plants to selection by herbivores can be quite complex and involve traits that allow plants to escape, defend or tolerate the fitness consequences of herbivore attack (Rausher 1992). Traits thought to offer protection against herbivore attack have received the most attention and include direct defences such as toxins, antifeedants and compounds which function as antinutrients (digestibility-reducing compounds; Fig. 1) as well as indirect defences that facilitate 'top-down' control of herbivores by the predators, pathogens and parasitoids of herbivores (Price 1984, Takabayashi

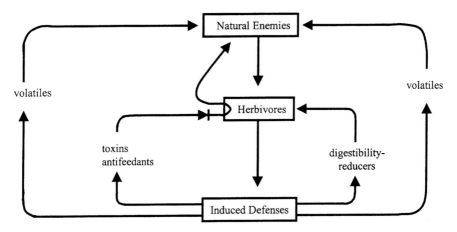

FIG. 1. The complexity of ecological interactions in a plant's community can undermine the function of an induced defence (e.g. when a toxin is sequestered by a plant's herbivores and used to protect the herbivore against its own natural enemies, thereby turning a plant defence against itself) or can be essential for its function (when a defence slows the growth rate of an herbivore, without which the third trophic level might cause a plant to lose more, rather than less, tissues to the herbivore). Volatile emissions from attacked plants, which are thought to increase the hunting efficiency of the natural enemies by making herbivores more 'apparent' to them, may function as an indirect defence.

& Dicke 1996, Turlings et al 1995). For example, some plants release volatiles when attacked and these volatiles are thought to guide parasitoids or other members of the third trophic level to the feeding herbivore (Turlings et al 1995, DeMoraes et al 1998; Fig. 1). Some direct defences, particularly those that slow herbivore growth by reducing their digestive efficiency, may not function as defences (i.e. do not increase the fitness of plants in environments with herbivores) without the third trophic level, because slow-growing herbivores may eat more leaf material to complete development than fast-growing herbivores (Jongsma & Bolter 1997, Moran & Hamilton 1980). Hence, for some direct defences we should expect coordinated induction of direct and indirect defences (Fig. 1). On the other hand, these higher-level ecological interactions can wreak havoc with the defensive function of other direct defences. For example, plant toxins are frequently sequestered by specialist herbivores for their own defence against their parasitoids and predators (Duffey et al 1986, Barbosa & Saunders 1985, Malcolm & Zalucki 1996), thereby turning a plant's defences against itself (Fig. 1). With such chemical defences, we might find use of direct defences to be incompatible with the deployment of indirect defences.

Many direct and most indirect defences are induced by herbivore attack and their inducible expression suggests that the selective advantage these traits provide are conditional, dependent on particular ecological circumstances. The most commonly invoked explanation for inducible expression is that although chemical defences are beneficial and increase a plant's fitness when it is under attack, they are costly when not needed, utilizing resources that could be used instead for growth or reproduction. Recent work has demonstrated that herbivore-induced direct defence responses can exact large resource demands on a plant's metabolism. For example, most of the large increases in respiration that occur after wounding can be attributed to the metabolic demands of induced furanocourmarin synthesis in wild parsnip (Zangerl et al 1997). If such large metabolic demands are commonly incurred when defences are activated, coordinated 'budget-balancing' alterations in metabolism are expected. Consistent with this expectation is the evidence that the jasmonate cascade, which plays an important role in activating many herbivore-induced responses, orchestrates both the up-regulation of defence genes and the down-regulation of many 'housekeeping' genes (Wasternack & Parthier 1997, Wasternack et al 1998).

In addition to the physiolgical costs of resistance, 'ecological' trade-offs, such as those between resistance and susceptibility to different natural enemies (Simms 1992, Adler & Karban 1994, Rausher 1996), mutualists (Euler & Baldwin 1996, Strauss 1997) or between resistance and tolerance to herbivory (van der Meijden et al 1988, Rosenthal & Kotanen 1994, Fineblum & Rausher 1995, Mauricio et al 1997) may result in significant fitness reductions for a well-defended plant, independently of the resource demands of the resistance traits. These ecological costs may be particularly important for some indirect defences, such as the release of volatile 'alarm' signals which are likely to make only small demands on a plant's metabolism (Dicke & Sabelis 1992). The defensive value of the release of volatile signals lies in its information content to the third trophic level and a plant has little control over the response of its ecological community to an increase in emission. Under some circumstances, a release of volatiles may increase a plant's apparency (*sensu* Feeny 1976) and attract its herbivores (Takabayashi & Dicke 1996). Since the plant no longer controls the signal once it has been released into the environment, the functional consequences for the plant of releasing volatiles are likely to be more unpredictable than they are for direct defences. For example, herbivore-induced volatile emissions from crab-apple appear to serve as host location cues for the Japanese beetle and contribute to the formation of large feeding aggregations of this pest species (Loughrin et al 1996). While the herbivore-induced release of volatiles has been demonstrated to attract parasitoids in laboratory settings, the ecological significance of these emissions for plants in nature remains largely unexplored.

These considerations suggest that while the inducible expression of both direct and indirect defences may be constrained to varying degrees by their physiological and 'ecological' costs, the fitness consequences of using either type of defence are likely to be highly dependent on the particular ecological situation a plant finds itself in. In this review, I consider the functional complexity involved in the use of both direct and indirect defences in the context of a well-characterized direct induced defence, the wound-induced production of nicotine in native *Nicotiana* species. I conclude that only with an intimate understanding of a plant's natural history can one hope to make sense of the complexity of plant responses to herbivory.

Wound- and *Manduca*-induced responses in native *Nicotiana* species

The production of the toxic alkaloid nicotine in two native species of *Nicotiana* (*N. sylvestris* and *N. attenuata*) illustrates the complexity of the mechanisms responsible for an induced direct defence, as well as the complexity of their ecological consequences. Both *Nicotiana* species are annuals and accumulate nicotine at 0.1–1.0% leaf dry mass in undamaged plants. After real or simulated folivory to plants grown in the glasshouse (Baldwin 1988a,b,c, 1989) or in natural populations (Baldwin & Ohnmeiss 1993, Baldwin 1998), leaf nicotine concentrations increase dramatically, typically fourfold, and under some circumstances 10-fold, to attain concentrations that are sufficient to deliver what is for many herbivores a lethal dose in a single meal and also protect induced tissues from herbivores that are adapted to nicotine (Baldwin 1991, 1988c). The within-plant distribution of nicotine is broadly consistent with the predictions of Optimal Defence theory (McKey 1979, Zangerl & Bazzaz 1992), which argues that defence metabolites are distributed within plants so that their fitness benefits are optimized in light of their fitness costs and are allocated preferentially to tissues with high fitness value to the plant and a high probability of attack. Young leaves, stems and reproductive parts tend to have the highest concentrations of nicotine, while roots and old leaves have the lowest (T. E. Ohnmeiss & I. T. Baldwin, unpublished work, Ohnmeiss et al 1997, Baldwin & Karb 1995, Euler & Baldwin 1996). While little is known about how the within-plant allocations come about, the induced response is clearly a whole-plant trait involving components distributed throughout the plant.

The results of many experiments support the model that nicotine is largely synthesized in the roots of both species and transported to the shoots in the xylem stream (reviewed in Baldwin 1999). The large spatial separation of the site of herbivore attack (shoots) and the site of nicotine synthesis (roots) requires a long-distance signal transduction cascade and the results of many experiments demonstrate that jasmonic acid (JA) is an essential component of this cascade (reviewed in Baldwin 1999). Our current working model for the long-distance signal transduction cascade is that wounding increases JA pools in shoots, which either directly through transport, or indirectly through a signal such as systemin (McGurl et al 1992), increases JA pools in roots; these, in turn, stimulate nicotine synthesis in the roots and increase nicotine pools throughout the plant (Fig. 2 insert).

From the above considerations, it is clear that induced nicotine production must be considered a whole-plant trait, since the response requires components distributed throughout the plant. Whole-plant nicotine induction is best described allometrically (by regressing the nicotine pool in the plant against the nicotine-free biomass acquired from a series of harvests) and the results of many such analyses have characterized the induced nicotine response in terms of allometrically determined set points for whole-plant nicotine pools (Baldwin 1997, Ohnmeiss & Baldwin 1994). These set points increase when a plant synthesizes a larger amount of nicotine per unit of growth in response to wounding or wound signals, but are not affected by nitrogen availability (Lynds & Baldwin 1998, Baldwin et al 1994a). The allometric slopes of uninduced rosette-stage plants tend to be slightly less than unity, which means that whole-plant nicotine concentrations decrease with growth (Ohnmeiss & Baldwin 1994). This decrease in whole-plant nicotine concentrations as plants mature reflects both the increasing proportion of older leaves on a plant which tend to have lower nicotine concentrations than younger leaves (Ohnmeiss et al 1997) and a waning of the rate of nicotine synthesis in the roots. As *N. sylvestris* plants begin to flower, they lose their ability to increase *de novo* nicotine production in response to leaf wounding (T. E. Ohnmeiss & I. T. Baldwin, unpublished work) but can redistribute the nicotine they already have in response to wounding (Fig. 2, left). This redistribution occurs principally towards the reproductive parts and young leaves. Diminished inducibility over development is due neither to the decreased ability of wounded leaves to produce JA after wounding nor to the decreased ability of the roots to respond with increased nicotine synthesis after exposure to jasmonates (T. E. Ohnmeiss & I. T. Baldwin, unpublished work). Rather, the lack of response appears to be due to a decreased ability to transport the wound signal from the leaf to the roots. Because reproductive tissues are strong sinks which 'attract' photoassimilates from source leaves, the sink-strength of roots wanes as plants flower, and these alterations in whole-plant source–sink relationships over ontogeny may be responsible for the waning of inducible nicotine production after flowering. Moreover, such an alteration may also explain the lack of inducibility in plants that have become 'pot-bound' (Baldwin 1988b) and are no longer strongly exporting photoassimilates from leaves to roots.

After understanding how *Nicotiana* responds to standardized mechanical wounding, we were interested in comparing the plant responses to damage from

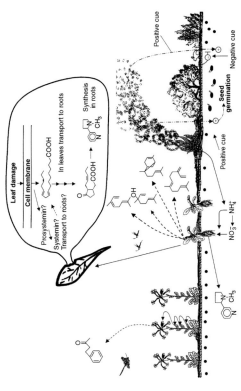

FIG. 2. The natural history of inducible nicotine production in *Nicotiana attenuata*. The plant is a herbaceous annual which initiates its germination from a long-lived seed bank in response to positive germination signals (+) produced by the combustion of cellulose during wildfires in sage-juniper habitats of the Great Basin desert and the pyrolytic removal of germination inhibitors (−) (negative cues) which leach from the accumulated litter in the A_0 horizon under the dominant vegetation. When rosette-stage plants are wounded or attacked by herbivores (rabbits or larvae of *Manduca* spp), the endogenous wound signal cascade producing jasmonic acid (JA) is activated in wounded tissues, increasing JA pools in shoots, which either directly through transport, or indirectly through a signal like systemin, increases JA pools in roots; these, in turn, stimulate nicotine synthesis in the roots and increase nicotine pools throughout the plant. Nicotine is preferentially synthesized from NH_4, which is the most abundant form of nitrogen available in the soil immediately after a fire and this preference may reduce the biosynthetic costs of nicotine production. As plants reach reproductive maturity, their ability to induce *de novo* nicotine synthesis wanes, but they reallocate previously synthesized nicotine from older leaves to younger leaves and reproductive parts in response to wounding. When larvae of *M. sexta* attack plants or their oral secretions are applied to leaves, the jasmonate cascade is dramatically amplified and attacked leaves release linalool, ocimene, bergamotene and farnesene, which may function to increase the apparency of the attacking herbivore to its parasitoids. The plants increase their apparency to nocturnal pollinators by releasing benzyl acetone from the outer lips of their corolla and withdrawing the nicotine into the base of their corolla below the nectar rewards and the attachment of the filaments. In short, the use of a highly toxic, nitrogen-intensive toxin has had a number of consequences for numerous other aspects of the plant's natural history.

the nicotine-adapted herbivore, *Manduca sexta*, with those to mechanical simulations of feeding damage. These comparisons revealed that plants respond differently to this herbivore than to the damage the herbivore causes. Larval feeding suppressed the nicotine response induced by a comparable amount of mechanical wounding (Baldwin 1988c, McCloud & Baldwin 1997) and, surprisingly, dramatically amplified the wound-induced production of JA (McCloud & Baldwin 1997). In addition, both of these alterations in the plant's response could be mimicked by the application of oral secretions from the caterpillars to a mechanical wound in the leaf (McCloud & Baldwin 1997). These adapted larvae could be feeding in a 'stealthy' fashion, reducing the nicotine responses below that of plants suffering a comparable amount of tissue loss from the same portions of their canopy over the same period and uncoupling the relationship between the endogenous wound signal, JA, and the response to this wound signal (Baldwin 1998, Ohnmeiss et al 1997). However, the opposite explanation, namely that the plant, rather than being manipulated by this specialized herbivore, could be optimizing its defence response to this nicotine-tolerant herbivore, is also possible. This hypothesis is suggested by a closer analysis of the plant responses to this herbivore.

Feeding by *M. sexta* larvae or the application of their oral secretions or jasmonates to leaves results in a dramatic increase in the release of both monoterpenes (ocimene and linalool) and sesquiterpenes (bergamotene and farnesene; Fig. 2) (J. Kahl, C. Preston, R. Halitschke, A. Lorenz & I. T. Baldwin, unpublished work). These volatiles are primarily emitted from the leaf on which the larvae are feeding and are only emitted during the day. When these same volatile compounds are also released by corn and cotton in response to herbivore attack, they are thought to function as alarm calls, attracting parasitoids (Turlings et al 1990, Takabayashi & Dicke 1996, De Moraes et al 1998) and they may be playing a similar role in *N. attenuata*. If these volatile emissions function as an indirect defence, the plant may be actively suppressing its nicotine response so as to avoid undermining the use of the third trophic level as a defence against this relatively nicotine-tolerant herbivore (Krischik et al 1988). Although *M. sexta* does not actively sequester nicotine, its haemolymph retains quantities of the toxin sufficient to cause substantial mortality of its parasitoid wasp, *Cotesia congregata* (Barbosa et al 1991). Thus although nicotine is not one of the commonly sequestered secondary metabolites, it may wreak havoc with the plant's ability to use the third trophic level as a defence and the suppression of the normal wound-induced nicotine response may minimize the potential incompatibility between direct and indirect defences in this species.

Herbivory induces a complex suite of changes in these native *Nicotiana* species, the details of which we are only beginning to appreciate. Folivory induces the systemic accumulation of nicotine which is coordinately expressed with increases in photosynthetic capacity (Baldwin & Ohnmeiss 1994a). The magnitude of the nicotine accumulations depend on the type of herbivore causing the damage (McCloud & Baldwin 1997). When the herbivore is *M. sexta*, the amount of nicotine accumulated is reduced and a number of other responses are activated, including the release of volatiles from attacked leaves and increases in proteinase inhibitors, phenolics and polyphenyl oxidases (N. van Dam, U. Schittko & I. T. Baldwin, unpublished work). The fact that many of these responses are induced by jasmonates provides researchers with the ability to examine the consequences of these responses in plants growing in nature in order to examine their fitness consequences.

Fitness consequences of jasmonate-induced responses in native *Nicotiana*

In an attempt to understand the fitness consequences of jasmonate-induced responses, we have quantified lifetime viable seed production of jasmonate-induced plants growing in environments of increasing ecological complexity: in hydroponic culture in the laboratory (Baldwin et al 1998), soil-based culture both with (van Dam & Baldwin 1998) and without (Lynds & Baldwin 1998, Baldwin et al 1998) intraspecific competition, field plantations and natural populations (Baldwin 1998). In some experiments we have attempted to limit the induced responses to the nicotine response by only treating root tissues with jasmonates. For plants growing in native populations induced with root jasmonate treatments and subjected to native herbivores (Baldwin 1998), the fitness benefit of jasmonate-induced nicotine production are readily detected. For example, in one population of 300 pairs of jasmonate-induced and control plants growing in a two-year-old burn (*N. attenuata* is a post-fire annual) with a high herbivore load, 33% of the control plants lost more than 40% of their leaf area to herbivores, compared to 1% for jasmonate-treated plants. Similar patterns were seen in populations that experienced more modest rates of herbivory; in these populations a fitness benefit could easily be detected in the lifetime viable seed production of plants. For 300 pairs of plants growing in a one-year-old burn, 241 pairs (80%) survived to produce seed; of these surviving pairs, jasmonate-treated plants matured on average 11% more viable seeds (6 capsules or a 349-viable-seed-plant^{-1} benefit of jasmonate treatment) than their untreated counterparts. Interestingly, this same field study also provided evidence that jasmonate-induced plants incurred a large fitness cost when they were not attacked by herbivores. When plants grew in populations where attack rates were very low because the plants were protected by fencing and insecticides, or when plant pairs simply escaped detection by the herbivore community, jasmonate-treated plants produced 17–26% fewer lifetime viable seed than their untreated partners (Baldwin 1998). These reductions correspond to the reduction associated with removing half a plant's leaf area at the rosette stage of growth, which reduces capsule production by 27% (Baldwin 1998). The fitness costs of jasmonate-induction observed in this experiment mirror the costs of wound-induction observed in a plantation experiment with the sibling species *N. sylvestris* (Baldwin et al 1990) as well as the fitness costs of plants growing in soil culture in the greenhouse (Baldwin et al 1998).

In summary, when plants are protected or escape from herbivore attack, eliciting wound responses with jasmonates significantly reduces the number of viable seeds produced at the time of senescence. This fitness cost, however, becomes a net fitness benefit when plants grow in environments with high rates of herbivore attack; here the fitness benefits of the responses induced by jasmonates outweigh the fitness costs. Many mechanisms are potentially responsible for the reduction of seed production. However, the similarity of the fitness costs of root jasmonate induction for plants growing in natural populations without herbivores with those observed in greenhouse-grown plants, argue that physiological rather than ecological processes are responsible for the reduction in seed production. When plants are induced and grown with an uninduced conspecific in the same container, the reduction in seed set associated with jasmonate induction is much greater than it is when plants are grown with a comparably-induced plant or without competition. Moreover, the uninduced competitor realizes a large 'opportunity fitness benefit' of growing adjacent to an induced plant (I. T. Baldwin & W. Hamilton, unpublished work, van Dam & Baldwin 1998). These results suggest that the physiological basis of the high fitness costs of jasmonate induction is a reduction in an induced plant's ability to compete for limiting resources with an uninduced competitor.

An examination of trade-offs between nicotine and seed production in a nitrogen currency is particularly germane for these plants. A molecule of nicotine contains two atoms of nitrogen and when a plant is induced, 6% of a plant's entire nitrogen content resides in nicotine (Baldwin et al 1998). Because this investment cannot be recouped by metabolism (Baldwin et al 1998, Baldwin & Ohnmeiss 1994b, Baldwin et al 1994a), the nitrogen remains unavailable for other activities such as seed production. This investment of nitrogen likely underestimates the actual costs, because it doesn't include the nitrogen incorporated into the nicotine biosynthetic enzymes or used in the transport and storage of nicotine. An investment of this magnitude will likely reduce seed production, given that nitrogen availability is demonstrably a fitness-limiting resource for these plants (van Dam & Baldwin 1998, I. T. Baldwin & W. Hamilton, unpublished work). Moreover, with ^{15}N labelling experiments, we have demonstrated that induced plants acquire less nitrogen than uninduced counterparts when grown in communial hydroponic chambers (I. T. Baldwin & W. Hamilton, unpublished work).

However, the nitrogen demands of nicotine induction do not necessarily result in a fitness cost because *N. attenuata* has evolved life history traits that mitigate the fitness consequences of this nitrogen-intensive defence (Fig. 2). It is a herbaceous annual which is an important but ephemeral component (usually for only three growing seasons), of the post-fire annual community in burned sagebrush, blackbrush and pinyon-juniper forests of the Great Basin desert (Baldwin et al 1994b, Baldwin & Morse 1994, Preston & Baldwin 1999). After fires it mass-germinates from long-lived seedbanks and its seeds are largely dormant until exposed to aqueous extracts of wood smoke. By synchronizing its germination with the post-fire environment, it takes advantage of the high-nitrogen soils that characterize such environments (Baldwin & Morse 1994, Lynds & Baldwin 1998). The majority of soil nitrogen after a fire exists as NH_4, which is rapidly oxidized to NO_3 during post-fire succession. *N. attenuata* preferentially uses NH_4 rather than NO_3 for nicotine production (Lynds & Baldwin 1998), perhaps because of the lower biosynthetic costs of synthesizing the alkaloid from the NH_4 (2.86 g glucose/g nicotine) compared to that of synthesizing it from NO_3 (3.62 g glucose/g nicotine (Gershenzon 1994). By producing dormant seeds that germinate after fires, this species selects environments with high soil NH_4 levels, which in turn allow it to defray the physiological costs of this resource-intensive chemical defence. This suggests that the selective importance of the nitrogen-based physiological costs will strongly depend on the germination behaviour of the seeds.

Future directions: functional consequences of the complexity of induced responses

We are only just beginning to appreciate the complexity of the changes that occur in a plant after herbivore attack (Bergey et al 1996) and it is clear that a detailed understanding of the natural history of the organism exhibiting conditional behaviours is essential to generating realistic, testable hypotheses about the fitness costs and benefits maintaining these plastic responses in nature. The fitness consequences of particular combinations of the different plastic responses will be determined by the particular ecological milieu that the plant finds itself in. Hence, an understanding of these fitness consequences will require an extensive amount of experimental fieldwork in which the components of the responses are independently manipulated. *Nicotiana* plants have both direct and indirect defences at their disposal and at this stage in our understanding, the following predictions about the functional interactions of these defences can be made:

(1) Indirect defences should be coordinately expressed with direct defences that slow the growth of herbivores and are not incompatible with the function of indirect defences (Fig. 1). Hence, as long as induced volatile emissions do not incur large 'ecological' costs by attracting herbivores, they should be

expressed in concert with the induced direct defences such as proteinase inhibitors and polyphenyl oxidases, which are likely to function as anti-nutritive defences. If direct defences employ toxins that can be sequestered by a specialist herbivore and used against the herbivore's predators and parasitoids, the use of these direct defences is incompatible with the use of indirect defences, and direct and indirect defences should not be deployed together. These predictions about the tailoring of direct and indirect defence responses presumes that plants are able to recognize attack by specialist herbivores.

(2)   The deployment of direct defences are likely to incur larger physiological costs than indirect defences. Therefore, when the fitness costs of direct defences are too large to provide a net fitness benefit, a plant should rely primarily on indirect defences. For example, when nicotine induction is particularly costly — due to low nitrogen supply rates in the environment or to the resource demands of flowering — plants should switch from using direct to indirect defences. However, when their fitness costs are not too great, plants should preferentially deploy direct defences over indirect defences due to their greater reliability as a defences.

Acknowledgments

Supported by the National Science (DEB-9505950) and the Max-Planck-Gesellschaft. Many of the ideas developed in this chapter benefited from discussion with C. Preston. I thank E. Claußen for drafting the figures and compiling the bibliography.

References

Adler FR, Karban R 1994 Defended fortresses or moving targets? Another model of inducible defenses inspired by military metaphors. Am Nat 144:813–832

Baldwin IT 1988a The alkaloidal reponses of wild tobacco to real and simulated herbivory. Oecologia 77:378–381

Baldwin IT 1988b Damage-induced alkaloids in tobacco: pot-bound plants are not inducible. J Chem Ecol 14:1113–1120

Baldwin IT 1988c Short-term damage-induced increases in tobacco alkaloids protect plants. Oecologia 75:367–370

Baldwin IT 1989 Mechanism of damaged-induced alkaloids in wild tobacco. J Chem Ecol 15:1661–1680

Baldwin IT 1991 Damage-induced alkaloids in wild tobacco, In: Raupp MJ, Tallamy DW (eds) Phytochemical induction by herbivores. Wiley, New York, p 47–49

Baldwin IT 1997 Allometric limits to the induced accumulation of nicotine in native tobacco. Plant Species Biol 11:107–114

Baldwin IT 1998 Jasmonate-induced responses are costly but benefit plants under attack in native populations. Proc Natl Acad Sci USA 95:8113–8118

Baldwin IT 1999 Inducible nicotine production in native *Nicotiana* as an example of adaptive phenotypic plasticity. J Chem Ecol 25:3–32

Baldwin IT, Karb MJ 1995 Plasticity in allocation of nicotine to reproductive parts in *Nicotiana attenuata*. J Chem Ecol 21:897–909

Baldwin IT, Morse L 1994 Up in smoke: II. Germination of *Nicotiana attenuata* in response to smoke-derived cues and nutrients in burned and unburned soils. J Chem Ecol 20:2373–2391

Baldwin IT, Ohnmeiss TE 1993 Alkaloidal responses to damage in *Nicotiana* native to North America. J Chem Ecol 19:1143–1153

Baldwin IT, Ohnmeiss TE 1994a Coordination of photosynthetic and alkaloidal responses to damage in uninducible and inducible *Nicotiana sylvestris*. Ecology 75:1003–1014

Baldwin IT, Ohnmeiss TE 1994b Swords into plowshares? *Nicotiana sylvestris* does not use nicotine as a nitrogen source under nitrogen-limited growth. Oecologia 98:385–392

Baldwin IT, Sims CL, Kean SE 1990 The reproductive consequences associated with inducible alkaloidal responses in wild tobacco. Ecology 71:252–262

Baldwin IT, Karb MJ, Ohnmeiss TE 1994a Allocation of ^{15}N from nitrate to nicotine: production and turnover of a damage-induced mobile defense. Ecology 75:1703–1713

Baldwin IT, Staszak-Kozinski L, Davidson R 1994b Up in smoke: I. Smoke-derived germination cues for the post-fire annual *Nicotiana attenuata* Torr Ex Watson. J Chem Ecol 20:2345–2371

Baldwin IT, Gorham D, Schmelz EA, Lewandowski C, Lynds GY 1998 Allocation of nitrogen to an inducible defense and seed production in *Nicotiana attenuata*. Oecologia 115: 541–552

Barbosa P, Saunders JA 1985 Plant allelochemicals: linkages between herbivores and their natural enemies. In: Cooper-Driver GA, Swain T, Conn EE (eds) Chemically mediated interactions between plants and other organisms. Plenum, New York (Recent Adv Phytochem 19) p 107–137

Barbosa P, Gross P, Kemper J 1991 Influence of plant allelochemicals on the tobacco hornworm and its parasitoid, *Cotesia congregata*. Ecology 72:1567–1575

Bergey DR, Hoi GA, Ryan CA 1996 Polypeptide signaling for plants defensive genes exhibits analogies to defense signaling in animals. Proc Natl Acad Sci USA 93:12053–12058

De Moraes CM, Lewis WJ, Paré PW, Alborn HT, Tumlinson JH 1998 Herbivore-infested plants selectively attract parasitoids. Nature 393:570–573

Dicke M, Sabelis MW 1992 Cost and benefits of chemical information conveyance: proximate and ultimate factors. In: Roitberg BD, Isman MB (eds). Insect chemical ecology: an evolutionary approach. Chapman & Hall, New York, p 122–155

Duffey SS, Bloem KA, Campbell BC 1986 Consequences of sequestration of plant natural products in plant–insect–parasitoid interactions. In: Boethel DJ, Eikenbarry RD (eds) Interactions of plant resistance and parasitoids and predators of insects. Ellis Horwood, Chichester, p 31–60

Euler M, Baldwin IT 1996 The chemistry of defense and apparency in the corollas of *Nicotiana attenuata*. Oecologia 107:102–112

Feeny P 1976 Plant apparency and chemical defense. In: Wallace JW, Mansell RL (eds) Biochemical interactions between plants and insects. Plenum, New York (Recent Adv Phytochem 10) p 1–40

Fineblum WL, Rausher MD 1995 Tradeoff between resistance and tolerance to herbivore damage in a morning glory. Nature 377:517–520

Gershenzon J 1994 The cost of plant chemical defense against herbivory: a biochemical perspective. In: Bernays EA (ed) Insect–plant interactions. CRC Press, Boca Raton, FL, p 105–173

Jongsma MA, Bolter C 1997 The adaptation of insects to plant protease inhibitors. J Insect Physiol 43:885–895

Krischik VA, Barbosa P, Reichelderfer C 1988 Three trophic level interactions: allelochemicals, *Manduca sexta* (L.) and *Bacillus thuringiensis* var. *kurstaki* Berliner. Environ Entomol 17: 476–482

Loughrin JH, Potter DA, Hamiltonkemp TR, Byers ME 1996 Role of feeding-induced plant volatiles in aggregative behavior of the japanese beetle (Coleoptera, Scarabaeidae). Environ Entomol 25:1188–1191

Lynds GY, Baldwin IT 1998 Fire, nitrogen, and defensive plasticity. Oecologia 115:531–540

Malcolm SB, Zalucki MP 1996 Milkweed latex and cardenolide induction may resolve the lethal plant defence paradox. Entomol Exp Appl 80:193–196

Mauricio R, Rausher MD, Burdick DS 1997 Variation in the defense strategies of plants: are resistance and tolerance mutually exclusive? Ecology 78:1301–1311

McCloud ES, Baldwin IT 1997 Herbivory and caterpillar regurgitants amplify the wound-induced increases in jasmonic acid but not nicotine in *Nicotiana sylvestris*. Planta 203:430–435

McGurl B, Pearce G, Orozco-Cardenas M, Ryan CA 1992 Structure, expression, and antisense inhibition of the systemin precursor gene. Science 255:1570–1573

McKey D 1979 The distribution of secondary compounds within plants. In: Rosenthal GA, Janzen DH (eds) Herbivores: their interaction with secondary plant metabolites. Academic Press, New York, p 56–133

Moran N, Hamilton WD 1980 Low nutritive quality as a defense against herbivores. J Theor Biol 86:247–254

Ohnmeiss TE, Baldwin IT 1994 The allometry of nitrogen allocation to growth and an inducible defense under nitrogen-limited growth. Ecology 75:995–1002

Ohnmeiss T, McCloud ES, Lynds GY, Baldwin IT 1997 Within-plant relationships among wounding, jasmonic acid, and nicotine: implications for defence in *Nicotiana sylvestris*. New Phytol 137:441–452

Preston CA, Baldwin IT 1999 Positive and negative signals regulate germination in the post-fire annual, *Nicotiana attenuata*. Ecology 80:481–494

Price PW 1984 Insect ecology, 2nd edn. Wiley, New York

Rausher MD 1992 Natural selection and the evolution of plant–insect interactions. In: Roitberg BD, Isman MB (eds) Insect chemical ecology: an evolutionary approach. Chapman & Hall, New York, p 20–88

Rausher MD 1996 Genetic analysis of coevolution between plants and their natural enemies. Trends Genet 12:212–217

Rosenthal JP, Kotanen PM 1994 Terrestrial plant tolerance to herbivory. Trends Ecol Evol 9:145–148

Simms EL 1992 Costs of plant resistance to herbivores. In: Fritz RS, Simms EL (eds) Plant resistance to herbivores and pathogens: ecology, evolution, and genetics. University of Chicago Press, Chicago, IL, p 392–425

Stahl E 1888 Pflanzen und Schnecken. Eine biologische Studie über die Schutzmittel der Pflanzen gegen Schneckenfrass. Jenaische Zeitschrift für Naturwissenschaft und Medizin 22:1–126

Strauss SY 1997 Floral characters link herbivores, pollinators, and plant fitness. Ecology 78:1640–1645

Takabayashi J, Dicke M 1996 Plant–carnivore mutualism through herbivore-induced carnivore attractants. Trends Plant Sci 1:109–113

Turlings TCJ, Tumlinson JH, Lewis WJ 1990 Exploitation of herbivore-induced plant odors by host-seeking parasitic wasps. Science 250:1251–1253

Turlings TCJ, Loughrin JH, McCall PJ, Röse USR, Lewis WJ, Tumlinson JH 1995 How caterpillar-damaged plants protect themselves by attracting parasitic wasps. Proc Natl Acad Sci USA 92:4169–4174

van Dam NM, Baldwin IT 1998 Costs of jasmonate-induced responses in plants competing for limited resources. Ecology Letters 1:30–33

van der Meijden E, Wijn M, Verkaar HJ 1988 Defense and regrowth, alternative plant strategies in the struggle against herbivores. Oikos 51:355–363

Wasternack C, Parthier B 1997 Jasmonate-signalled plant gene expression. Trends Plant Sci 2:302–307

Wasternack C, Ortel B, Miersch O et al 1998 Diversity in octadecanoid-induced gene expression of tomato. J Plant Physiol 152:345–352

Zangerl AR, Bazzaz FA 1992 Theory and pattern in plant defense allocation. In: Fritz RS, Simms EL (eds) Plant resistance to herbivores and pathogens: ecology, evolution, and genetics. University of Chicago Press, Chicago, IL, p 363–391

Zangerl AZ, Arntz AM, Berenbaum MR 1997 Physiological price of an induced chemical defense — photosynthesis, respiration, biosynthesis, and growth. Oecologia 109:433–441

DISCUSSION

Tumlinson: Did you say that when the plant is flowering it does not produce volatiles in response to herbivory?

Baldwin: No, when a plant is flowering it does not induce nicotine in response to herbivory. This is a cost-sensitive time for the plant; a time when the plant may want to use a cheaper form of defence, and volatile production may fit this bill.

Schultz: Can you induce any of the other defences, such as proteinase inhibitors or phenolics, when the plants are flowering?

Baldwin: It is too early to say for proteinase inhibitors, but induced volatile emissions appear to be higher during flowering.

Schultz: We're trying to induce a carbon cost in tobacco. We want to work with chloroplast transport mutants to see if we can figure out where the carbon is coming from when demand for it goes up. (We wouldn't have worked with tobacco for carbon-based defences except that it has the necessary chloroplast carbon transport mutants and we can't get those any place else.) We are looking for a situation in which we can induce only terpenoid or phenolic synthesis in tobacco.

Firn: Is it the flowering or the induction process in the leaves which change the physiology of the plant? If you take the developing seed off, what happens?

Baldwin: If you take the seed off, you can get the plants to induce nicotine again, so it seems that it is a hormonal signal from the seed.

Schultz: Are you looking for glucose oxidase in saliva?

Baldwin: Glucose oxidase seems to function primarily by producing hydrogen peroxide. It does not activate the JA cascade in the way that spit does. Spit's amplification of the JA cascade is rather dramatic: we get at least a fourfold increase above that produced by wounding. Hydrogen peroxide doesn't do anything close to that. We treat our spit by boiling and filtering: few lytic enzymes would survive that treatment.

Schultz: I was really interested in your ecological interpretation of suppression. We have a caterpillar that fails to induce oak trees, and evidently suppresses condensed tannin synthesis. I thought, 'Oh great, caterpillar beats tree'. But as we found with the gypsy moth, the biggest impact of high tannin content in leaves is going to be to protect the caterpillar against disease. If the plant is suppressing tannin synthesis, it is increasing the risk of disease in the insect, and the difference between it and the gypsy moth might be that the plant can't do that; it doesn't recognize the gypsy moth.

Scheel: How specific is the perception of the signals from spit, with respect to plant species? Do many plant species recognize the same signals?

Baldwin: The trouble is, how do you know that a plant recognizes something? It takes a lot of work to figure out what a plant is responding to. The best we can do is to take spit from other species and apply it to a plant for which we know something about its recognition responses. We've only studied one example of this, and I wouldn't dare to extrapolate from this.

Tumlinson: In the work that Consuelo De Moraes did, in which she was looking at the response of plants to *Heliothis virescens* and *Helicoverpa zea*, she investigated corn, cotton and tobacco plants and there was a difference in the response of the plants to the two different herbivore species in every case (De Moraes et al 1998). *H. zea* doesn't normally feed on tobacco and *H. virescens* doesn't normally feed on corn; they both feed on cotton. The tobacco plant does respond to *H. zea* spit (or to *H. zea* feeding on it), and the corn responds to *H. virescens*, but not in the same way as they respond to other species.

Baldwin: I think the devil is in the details. It's not just a question of whether or not the plant responds. You may see a jasmonate burst and there may be some recognition component, but the crucial question is whether this actually translates to changes in transcription.

Takabayashi: How about the feeding pattern of the caterpillar? When the caterpillar starts feeding, each bite removes a certain amount of leaf tissue and the elicitor is applied along the damaged edge of the leaf at the same time. However, the following bite will remove the tissue to which the elicitor was applied in the previous bite. In this way, very little elicitor will be left along the edge of the leaf on which the caterpillar has fed. This kind of application may have a different effect from the treatment with jasmonic acid.

Baldwin: Manduca sexta larvae start out small and are slow, continual munchers. After feeding on a particular leaf for a couple of days they get up and find an uninduced leaf. If you induce all the leaves ahead of them they will continue to search, eventually moving to another plant. This is a good example that the best defence for a plant is not to kill the caterpillar, but to walk it on to your neighbour and turn it into an agent of competition. This may be what many induced defences are all about. But to answer your question, when *Manduca* goes into the third instar, it changes its feeding behaviour. They munch like crazy for an hour, and then go to sleep for about eight hours. Then they eat and sleep again. This has been interpreted largely as a response to dealing with how they fill their gut and empty it. One functional consequence of this rapid but episodic feeding mode, is that they could be eating up the wound signal their feeding damage causes before it can get out of the plant. At the later instars, they can eat fast enough — we have done the calculations and this is definitely possible. The jasmonate signal produced by a single bout of mechanical wounding takes about an hour to leave the leaf, during which time a caterpillar can eat a lot of leaf.

Haukioja: How sensitive is *Manduca* to nicotine? There are no rabbits in the environment, so does this nicotine production protect the plant?

Baldwin: It's hard for me to answer that with precision. In the laboratory, *Manduca sexta* does not care much about nicotine. High, induced concentrations slow growth rates only by a few percent. However, when you induce plants in the field with root treatment of methyl jasmonate, they're dramatically protected from herbivore attack. The question of how important nicotine is for this protection is not going to be answered with much precision until we can surgically inhibit nicotine biosynthesis without influencing a lot of other things.

Poppy: You mentioned that the caterpillar will walk around the plant trying to find non-induced leaves. Is there any suggestion that plants next to heavily induced plants will induce themselves? If the roots of a plant come into contact with the roots of an induced neighbour, is there any induction?

Baldwin: I spent the first year of my graduate degree looking for such an effect, and I saw no evidence of between-tobacco-plant induction.

Bohlmann: I have a question about the induced volatile formation. How does the level of induced emission compare to the constitutive level, and how much plant to plant variation is there in comparison?

Baldwin: For farnesene and bergamotene there is about a fourfold increase from a very low constitutive level. Linalool is variable; we have some lines that induce it and some that don't. The variation in volatile emission is rather impressive: some lines do it consistently for some time and then seem not to do it for a while.

Lamb: Are there particular organs or cell types that produce more volatiles than others? Is there any architecture to this?

Gershenzon: There is no information about that for herbivore-induced volatile production. In the initial experiments that we've done, we have been unable to localize the synthesis or release of volatiles to particular parts of the leaf. For maize, we compared the base, middle and tip of damaged leaves, but there's no indication of any differences between these regions in volatile production.

Tumlinson: We've looked at glandless varieties of cotton, and these produce essentially the same induced volatiles as glanded varieties of cotton.

Pickett: As genes are cloned for enzymes in these pathways, then obviously the sites of expression can be located and the associated promoter systems investigated.

Gershenzon: For specialized tissues that produce volatile compounds constitutively, such as resin blisters in fir (the system Jörg Bohlmann has worked with), the glandular trichomes in mint or floral tissues, it has been possible to localize biosynthetic genes involved in volatile production to specific cells and even particular subcellular compartments. This is not yet true for herbivore-induced volatiles, but it certainly would be useful to know which part of the plant is responding. We did do a quick experiment in cotton to see whether herbivore damage would cause responses in the apical meristem, often a site of constitutive volatile production. Cotton is a nice plant for such experiments because damage apparently leads to a systemic induction of volatiles (Röse et al 1996), and damage caused by herbivores increases the non-volatile terpenoid aldehyde feeding deterrents, especially in the apical meristem and in the youngest leaves (McAuslane & Alborn 1998). But the results of this experiment didn't indicate the localization of induced volatile production in the apical meristem.

Lamb: When an insect or a caterpillar is on a plant, what is its behaviour? Does it always eat on the leaf it is on, or does it sometimes pass across a leaf without attempting to eat it? In a plant that's induced for protection, does the herbivore just dislike the plant and go away from it or does it try to eat and then go away from it?

Baldwin: It samples the plant. The sampling may be very subtle: Paul Feeny has shown that a lot of insects smell with their feet, and tarsal tapping tells them something about the internal chemistry of the leaf.

Lamb: In all the discussion here, the induction seems to me to be a grazing response to actual mechanical damage. But is there any evidence that the plant just senses the presence of an insect without grazing? If the insect was prevented from chewing, would the plant know that it was there without the grazing?

Baldwin: I know of no published work demonstrating such an effect.

Pickett: Does anyone else wish to make a contribution on pre-feeding interactions with plants? We have mentioned egg laying — there are publications by Miriam Rothschild (Bergström et al 1995) and Anton Blaakmeer (Blaakmeer et al 1994a,b).

Baldwin: There's also a nice study from Monika Hilker's lab with elm leaf beetles, in which she's shown that ovipostion, which is an intraleaf event, induces volatile emissions (Meiners & Hilker 1997).

Pickett: With *Pieris brassicae* it is probably a superficial event, i.e. the use of glue to stick the eggs to the leaf. What about touch?

Gershenzon: There is John Steffen's work on tomato trichomes (Castanera et al 1996). When insects damage trichomes containing high levels of polyphenol oxidase (PPO), there is a reaction between phenolics and the PPO which causes the phenolics to polymerize and prevent the mouthparts from functioning. This is a rapid response, with no apparent *de novo* synthesis of defensive chemicals, but is something that can happen when an insect lands on the plant without actually feeding. Thus, perhaps other induced responses are triggered by simple contact.

Firn: Surely if this response is just based on touch then plants are going to have a problem with wind induction.

Schultz: We have done some work with this, and certainly bean plants acclimate after a while. Over a period of days you get responses and then they level off.

Karban: Induced responses to oviposition can be very effective, even if they are slow, as happens when 17 year cicadas oviposit into tree branches (White 1981). Trees that are oviposited on by cicadas respond to the eggs, and since these insects are periodical and highly synchronized, they feed for a 17 year period but only attack the tree as adults once every 17 years. A tree that is able to kill the eggs gets rid of potential feeders for a full 17 years. Even though its response — killings these eggs — takes place very slowly, it gets rid of the potential feeders for the next 17 years.

Dicke: With potato plants we measured the effect on Colorado potato beetles. When we had plants that were not attractive in a wind tunnel and we started shaking them, they became attractive (Bolter et al 1997).

Tumlinson: When Consuelo De Moraes took potted tobacco plants into the field, it took a while after they were set up in the field before they started attracting wasps. Shaking the plants appeared to shut down the volatile production. I've read that wind stimulation in certain cereals makes the plant put more resources into strengthening the stem and shuts down other processes.

Schultz: That's a huge literature. In our case, we found that in bean plants shaken by wind, peroxidase and PPO activity increases, as do all the enzymes responsible for lignification. You can even get this effect by exposing the plants to continuous low frequency sound — music.

Pickett: Does anyone have a view as to how this signal mechanism is operating?

Lamb: The reason I asked about this is that we've been studying the oxidative burst in relation to systemic acquired resistance, looking for hydrogen peroxide with various cytochemical agents. We noticed that if you devise conditions where you avoid photorespiration you can get leaves with very low background peroxide, except occasionally we saw a trichome really light up. Because trichomes stick out I thought this must be a mechanosensing phenomenon. We then took a soft paintbrush and stroked the trichomes, which locally sets up a huge burst: every trichome fires.

Schultz: One can induce an oxidative burst by squeezing individual cells between glass plates. One can really see this in studies of induction in cell cultures; the shape of the induction curves reflects the timing of culture transfer and the initiation of agitation, bouncing the cells off the container walls. It is an experimental problem in those systems.

Lamb: I see this as a biologically relevant phenomenon, because if something is moving across the leaf, whether it's wind, water or an insect, this is important information for the plant.

Karban: Part of this depends on what we're using as a control. If you have as your control tomato plants that are completely protected in the greenhouse, and compare those to plants that are handled a bit, the handled plants have higher levels of proteinase inhibitors. However, if you take plants out in the field, you can still get an increase in those proteinase inhibitors relative to controls that you haven't allowed herbivores on but are still getting by buffeted by everything that's out there.

Pickett: Chris Lamb, what is the mechanism for generating hydrogen peroxide in the case of the trichomes?

Lamb: We've done some pharmacological experiments that are technically quite tricky. The data are at least consistent with it being the NADPH oxidase. We have some antisense suppressed plants for NADPH oxidase, but I haven't yet persuaded this person to spend 10 minutes brushing the trichomes. I would really like this to be the case because, apart from anything else, for mutant screens this is fairly straightforward.

Firn: Tony Trewavas has shown intercellular calcium changing within seconds after exposing a plant to wind (Trewavas & Knight 1994). There's also a series of touch-sensitive genes known in plants (Braam & Davis 1990). The trouble is knowing how this kind of mechanical information benefits plants growing out in an environment where there are frequent periods of wind.

Scheel: Jack Schultz, with the bean plants do you see a refractory period after you have shaken them?

Schultz: Yes. This is only one contribution to the resistant status of the plant in the field. I don't have any kind of data set that would allow me to assign a fraction of variance among individuals to this source. And as Rick Karban is pointing out, for at least some plants that are regularly bent or mangled, the fact is that there's a high background against which you can still see insect-induced changes, so this is only part of the story. On the other hand, ecologists do report low densities of herbivorous insects in sites that are constantly wind-buffeted. The common interpretations are that the wind blows all insects away, or that the leaves are too tough or something. But nobody has systematically looked at those plants to see if they have developed an ongoing resistance because of the motion. We are supposed to do that experiment this summer with juvenile oaks and a series of wind regimes.

Scheel: There are many more different signal transduction components that are activated by the different stresses of touch, wind and wounding. One difference depending on the stress you give to a plant is the length and intensity of activation of signalling pathways. This sort of information may be involved in signalling for specific responses.

Schultz: One can collect a rich literature of cross effects to illustrate that. Plants exposed to ozone are more resistant to rust. The same plant infected by rust is more tolerant of ozone. We hypothesize that the signal transduction pathways have enough crossover points that there could be significant ecological outcomes. I think this is one of the big questions of this field: how many crossover points are there and do they have an ecological consequence? I started out extremely enthusiastic that there were lots of crossovers, but the more we look the more unique responses we find.

In the discussion after Lester Wadham's paper (p 71), I asked whether there are plant responses to aphids, such as hypersensitive responses, that might suggest a microbial element to aphid attack on plants. Evidently, some components of systemic required resistance, in particular the production of certain pathogenesis-related (PR) proteins, are indeed induced by aphids that are infected by endosymbiotic bacteria. So the plant response to aphids resembles a response to pathogens, as opposed to an insect.

Takabayashi: A hypersensitive-like response can be seen the spider mite–plant interaction. When plants are infested by Kanzawa spider mites, the infested region becomes necrotic. But this does not happen when a plant is infested by two-spotted spider mites.

Schultz: Another system like that is grape vines and phylloxera. One characteristic of resistant grape genotypes is an evident hypersensitive response when the insect tries to form a gall. But this is not data, it is deduced from a hand-drawn picture from the 1930s.

Baldwin: I have a comment about how to interpret this sort of evidence and whether or not it says anything about it being a pathogen-like response. The hypersensitive response is just a phenomenological statement about localized cell death, and recent work has demonstrated that there are many different mechanisms responsible for the hypersensitive response. Just because hypersensitive responses were first noticed in the pathogen literature, it does not mean that they are specific pathogen responses.

Schultz: I agree. The reason I raised this is that it has been postulated that this a common induced reaction to galling insects by plants. The argument in the literature is that it works because the galler can't go anywhere else, so it is a way for the plant to rid itself of the galling insect.

Firn: Do galling insects induce volatiles?

Wadhams: Not a lot, although we need to reexamine this with some of our newer techniques.

Schultz: They induce ethylene, peroxidases and PPOs, or they turn them down depending on the species.

References

Bergström G, Rothschild M, Groth M, Crighton C 1995 Oviposition by butterflies on young leaves: investigation of leaf volatiles. Chemoecology 6:147–158

Blaakmeer A, Stork A, van Veldhuizen A et al 1994a Isolation, identification, and synthesis of
    miriamides, new hostmarkers from eggs of *Pieris brassicae*. J Nat Prod (Lloydia) 57:90–99
Blaakmeer A, Hagenbeek D, van Beek TA, de Groot A, Schoonhoven LM, van Loon JJA 1994b
    Plant response to eggs vs. host marking pheromone as factors inhibiting oviposition by *Pieris
    brassicae*. J Chem Ecol 20:1657–1665
Bolter CJ, Dicke M, van Loon JA, Visser JH, Posthumus MA 1997 Attraction of Colorado
    potato beetle to herbivore damaged plants during herbivory and after its termination. J
    Chem Ecol 23:1003–1023
Braam J, Davis RW 1990 Rain-, wind-, and touch-induced expression of calmodulin and
    calmodulin-related genes in Arabidopsis. Cell 60:357–364
Castanera P, Steffens JC, Tingey WM 1996 Biological performance of Colorado potato beetle
    larvae on potato genotypes with differing levels of polyphenol oxidase. J Chem Ecol 22:91–101
De Moraes CM, Lewis WJ, Paré PW, Alborn HT, Tumlinson JH 1998 Herbivore-infested
    plants selectively attract parasitoids. Nature 393:570–573
McAuslane HJ, Alborn HT 1998 Systemic induction of allelochemicals in glanded and glandless
    isogenic cotton by *Spodoptera exigua* feeding. J Chem Ecol 24:399–416
Meiners T, Hilker M 1997 Host location in *Oomyzus gallerucae* (Hymenoptera: Eulophidae), an
    egg parasitoid of the elm leaf beetle *Xanthogaleruca luteola* (Coleoptera: Chrysomelidae).
    Oecologia 112:87–93
Röse USR, Manukian A, Heath RR, Tumlinson JH 1996 Volatile semiochemicals released from
    undamaged cotton leaves: a systemic response of living plants to caterpillar damage. Plant
    Physiol 111:487–495
Trewavas AJ, Knight MR 1994 Mechanical signalling, calcium and plant form. Plant Mol Biol
    26:205–214
White J 1981 Flagging: host defences versus oviposition strategies in periodical cicadas
    (*Magicicada* spp., Cicadidae, Homoptera). Can Entomol 113:727–738

Plant production of volatile semiochemicals in response to insect-derived elicitors

J. H. Tumlinson, P. W. Paré and W. J. Lewis*

*Center for Medical, Agricultural, and Veterinary Entomology, United States Department of Agriculture, Agricultural Research Service, 1700 Southwest 23rd Drive, Gainesville, FL 32608 and *Insect Biology and Population Management Research Laboratory, United States Department of Agriculture, Agricultural Research Service, Tifton, GA 31793, USA*

Abstract. An increase in the release of volatile compounds by plants in response to insect feeding is triggered by interaction of elicitors in the oral secretions of insect herbivores with damaged plant tissues. This herbivore damage triggers *de novo* biosynthesis of volatile plant metabolites derived from several different biochemical pathways. Natural enemies of herbivores use these volatile semiochemicals to locate their hosts. Although some volatile compounds are released from storage in plants immediately whenever damage to cells or glands occurs, the induced compounds are only synthesized and released during the light period. This often results in a delay between feeding damage and release of volatiles. Plants release the induced compounds from undamaged as well as damaged leaves. Thus, damage to only a few leaves results in a systemic response and release of volatiles by the entire plant. We propose that plants respond differently to individual herbivore species at least in part due to the composition of insect elicitors that come in contact with the plant. Specialist parasitoids can differentiate the volatile blends released due to damage by hosts from those resulting from non-host damage as well as from mechanical damage, thereby facilitating host location for the parasitoid. Elicitors in the oral secretions of beet armyworm caterpillars have been identified and synthesized.

1999 Insect–plant interactions and induced plant defence. Wiley, Chichester (Novartis Foundation Symposium 223) p 95–109

Plants, when attacked by insect herbivores, release volatile compounds that are used as cues by natural enemies of the herbivores to locate their hosts or prey. These chemically mediated tritrophic interactions have now been well documented in several agroecosystems. For example, Dicke and co-workers have shown that leaves from both lima bean (Dicke et al 1990) and cucumber (Takabayashi et al 1994a) plants as well as the foliage of apple trees (Takabayashi et al 1991) attacked by phytophagous mites release volatiles that attract predatory mites. Turlings et al (1990, 1991, 1993) demonstrated that corn seedlings damaged by beet armyworm larvae release a blend of terpenoids and indole that is attractive to *Cotesia marginiventris*, a parasitoid that attacks larvae of several species of Lepidoptera. Tritrophic systems such as these are of great interest because of the need to develop effective methods for biological control of insect pests of agricultural crops. Only by understanding these systems will we be able to make a fundamental shift in pest management practices from application of pesticides or other agents to kill the pests to an approach that promotes naturally occurring biological agents and other inherent strengths of agricultural ecosystems to keep pests within acceptable bounds (Lewis et al 1997).

Plant biosynthesis and release of volatiles in response to insect herbivory

Plants respond to insect feeding damage by releasing compounds synthesized by at least four different pathways. For example, the compounds released by cotton include indole, monoterpenes, sesquiterpenes, and the 'green leafy' hexenols and hexenals derived from the degradation of linolenic acid by lipoxygenases (Fig. 1). The compounds released vary with different plant species and even different varieties of the same species (Loughrin et al 1995, Takabayashi et al 1994a, Turlings et al 1998). However, in all systems studied, insect feeding damage elicits release of greater quantities and different blends of volatiles than does mechanical damage alone (Turlings et al 1990, Takabayashi et al 1994b, Paré & Tumlinson 1997a). Typically, application of insect herbivore oral secretion to damaged tissues of plants results in release of volatiles in similar quantities and proportions to those released in response to insect feeding (Turlings et al 1990, Paré & Tumlinson 1997a).

When corn seedlings are damaged by beet armyworm larval feeding during late evening or the dark period, only the hexenals and hexenols, products of the lipoxygenase degradation of linolenic acid, are released immediately. The release of other volatile compounds, terpenoids and indole, is delayed until the next light period (Turlings et al 1990). However, when the same insect species feeds on glanded cotton, a dichotomous volatile release pattern results (Loughrin et al 1994). Several monoterpenes and sesquiterpenes are released, along with the lipoxygenase products immediately after damage. Another set of terpenoids and indole follow a diurnal pattern of release with the maximum amounts released near the middle of the next photophase. Thus, it was proposed that the volatiles released by beet armyworm-damaged cotton consist of constitutive compounds, passively released at any time that mechanical damage to glands occurs, and induced compounds, which are only released during the photophase, presumably as a result of biochemical processes. It has also been noted that the release of

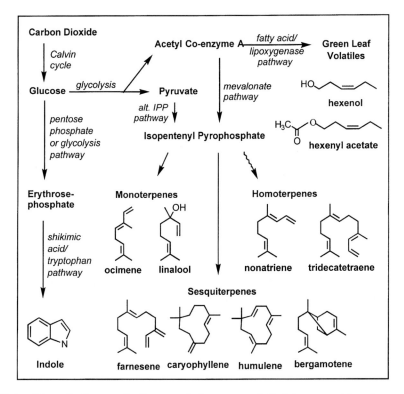

FIG. 1. Biosynthetic routes leading to volatile compounds (indole, terpenoids and green leaf volatiles) released by plants damaged by insect herbivores.

induced compounds by both cotton and corn increases with increased light intensity (J. H. Loughrin & J. H. Tumlinson, unpublished results). A similar increase in amounts of volatile compounds when lima bean leaves were held under higher light intensity was reported by Takabayashi et al (1994b). This suggests that the release of induced volatiles is driven by photosynthesis and that the induced compounds are synthesized *de novo*.

De novo biosynthesis of volatiles

One proposed mechanism for this delay is that plants store certain volatile compounds as glycosides which are cleaved by glycosidases present in insect saliva resulting in volatile release (Boland et al 1992). In fact, β-glucosidase activity has been measured in the regurgitant of *Pieris brassicae*, a herbivore that feeds on cabbage, and leaves treated with a commercial β-glucosidase released a volatile blend similar to that of leaves treated with *P. brassicae* regurgitant (Mattiacci et al 1995).

In experiments in which cotton plants were exposed to $^{13}CO_2$ continuously for 36 h (beginning 2 h after the start of continuous beet armyworm feeding) and volatiles were collected at 3 h intervals, we found that terpene biosynthesis and/or release in response to insect feeding damage occurs by more than one route. Several compounds incorporated substantial levels of ^{13}C within 30 h of exposure to $^{13}CO_2$ and were clearly synthesized *de novo* in response to insect damage. These highly labelled compounds included: (*E*)-β-ocimene (92% ^{13}C enrichment); linalool (86% ^{13}C enrichment); (*E,E*)-α-farnesene (80% ^{13}C enrichment); (*E*)-β-farnesene (99% ^{13}C enrichment); and the homoterpenes (*E*)-4,8-dimethyl-1,3,7-nonatriene (93% ^{13}C enrichment) and (*E,E*)-4,8,12-trimethyl-1,3,7,11-tridecatetraene (54% ^{13}C enrichment). During the same 30 h exposure to $^{13}CO_2$ an almost equal number of volatile terpenes incorporated very low levels or non-detectable levels of ^{13}C label. Compounds with low enrichment levels included: α-pinene (5% ^{13}C enrichment), β-pinene (5% ^{13}C enrichment), (*E*)-β-caryophyllene (5% ^{13}C enrichment); and α-humulene (5% ^{13}C enrichment). These volatiles were either released from storage, and thus not synthesized in response to insect wounding, or were synthesized from other precursor pools and thus did not incorporate the labelled carbon. The monoterpenes limonene and myrcene contained intermediate levels of ^{13}C label after 30 h of $^{13}CO_2$ exposure with 20% and 35% ^{13}C incorporation, respectively.

In a second experiment, plants were exposed to a 2 h pulse of $^{13}CO_2$, between 0900 and 1100 h, on the day after feeding damage began. The highly labelled terpenes rapidly incorporated ^{13}C, as indicated by the high percentage of labelled compounds in the volatiles released during the 3 h collection period overlapping the $^{13}CO_2$ pulse (Fig. 2). Furthermore, the rapid disappearance of the carbon label from terpenes collected during the subsequent 3 h periods on that day indicated a high rate of turnover of the highly labelled compounds. Greater than half of the ^{13}C was displaced from the labelled terpenes within the first 3 h collection period after $^{13}CO_2$ was removed.

The $^{13}CO_2$ labelling experiments clearly establish that in cotton the induced compounds are synthesized *de novo,* and rapidly released into the atmosphere after synthesis. However, in the cotton varieties we investigated, which were all glanded, several mono- and sesquiterpenes were not labelled, indicating they were released from storage. In corn, all of the compounds except for the lipoxygenase products appear to be induced. We have not yet determined whether or not they are synthesized *de novo*, but since they are only released during the photophase and the quantities released increase with increased light intensity we can speculate that they are also synthesized *de novo*.

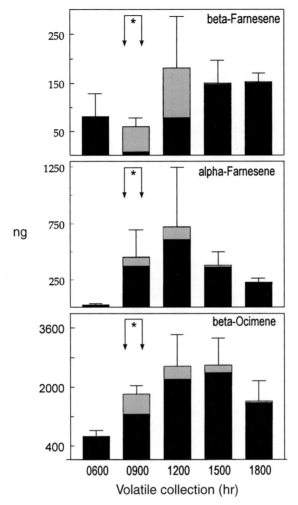

FIG. 2. The release and incorporation of ^{13}C label into cotton volatiles while beet armyworm caterpillars are allowed to feed on the plant. The arrows with an asterisk demarcate the interval in which the plants were exposed to synthetic air highly enriched with $^{13}CO_2$. The bars represent the total amounts of the indicated volatile compound collected during a 3 h interval, which began at the labelled time; grey portions of the bars represent the fraction of the collection that had incorporated ^{13}C label. Error bars represent SE of the total volatile emission.

Systemic biosynthesis and release of volatiles

In addition to the release of volatiles at the site of herbivore feeding, analysis of volatile emissions from unharmed leaves of insect-damaged plants has established that there is a systemic response. In both corn (Turlings & Tumlinson 1992) and cotton (Röse et al 1996), leaves distal to the site of herbivore feeding showed an increase in the release of volatiles. The chemical blend of volatiles from undamaged cotton leaves differs from the volatiles collected from the entire plant (Röse et al 1996). The products of the lipoxygenase pathway, including the hexenals and hexenols, which are released from freshly cut or damaged tissue, are not detected in the systemically released volatiles, except for (Z)-3-hexenyl acetate. In cotton, some of the monoterpenes and sesquiterpenes, as well as indole and isomeric hexenyl butyrates and 2-methyl butyrates, are only released from damaged leaves (Röse et al 1996). The terpenoids that are synthesized *de novo* in cotton leaves in response to herbivore damage (Paré & Tumlinson 1997a,b) are also released systemically from undamaged leaves of a herbivore-injured plant. The terpenoids that did not incorporate ^{13}C when damaged cotton plants were exposed to $^{13}CO_2$ were not released systemically.

These results raised the question of whether compounds released systemically are synthesized *de novo* in the leaves from which they are released, or are synthesized in the damaged leaves and transported to the undamaged leaves. To answer this question we collected volatiles from leaves in the upper portion of plants, after the bottom leaves of the plants, which were outside the volatile collection chamber, had been damaged by beet armyworm feeding for 45 h. During volatile collection the upper leaves were exposed to air containing $^{13}CO_2$. Analysis of the systemically released volatiles by gas chromatography–mass spectroscopy (GC–MS) showed that a high level of ^{13}C was rapidly incorporated into these compounds, indicating that they are synthesized *de novo* at the site of release (Paré & Tumlinson 1998). This suggests that a signal is transmitted by either volicitin (Alborn et al 1997) or another messenger from the site of damage to distal, undamaged leaves to trigger synthesis and release of volatile compounds. The transfer of such a signal to undamaged portions of the plant serves as a mechanism for amplifying the message that the plant is under herbivore attack.

Variability and specificity in the volatiles released by plants

At the species level, head space volatiles of several crop plants have been analysed with and without insect damage. Though the chemical blend as well as the total volatile release varies, there are some compounds that many species have in common. Repeatedly, the acyclic C_{11} homoterpene (E)-4,8-dimethyl-1,3,7-nonatriene and the C_{16} homoterpene (3E,7E)-4,8,12-trimethyl-1,3,7,11-tridecatetraene have been identified in the head space of herbivore-infested plants, including lima bean, apple, cowpea, cucumber, corn and cotton (Dicke 1994, Loughrin et al 1994, Turlings et al 1991).

Data on variation in volatiles released at the cultivar level is limited. In cotton, no significant differences were reported in the volatiles released among commercial cotton varieties, though a naturalized cotton variety did emit substantially greater quantities of volatiles per plant than any of the commercial hybrids (Loughrin et al 1995). Five varieties of corn exposed to herbivore injury differed significantly in the quantity and variety of terpenes released. The genotypes showed variation in their resistance to the important corn pest, fall armyworm as well. However these differences did not appear to affect the ability of the parasitoids *Cotesia marginiventris* and *Microplitis croceipes* to locate their host, the fall armyworm (Turlings et al 1995). Clearly the volatile blend released by an individual plant can vary somewhat without eliminating transmission of the signal.

Differences in the amount of volatiles released between individual plants are often subtle and can easily be masked by variations in environmental conditions that influence the plants' physiology. Several species including corn, cotton (J. H. Loughrin & J. H. Tumlinson, unpublished data) and lima bean (Takabayashi et al 1994b) respond to reduced light, due to either lower light intensity or shorter day length, with a decline in the release of herbivore-induced volatiles. On the basis of studies with lima bean, water stress seems to be directly related to volatiles released. With less water available for the plant, elevated levels of volatiles are released from infested individuals relative to non-water-stressed controls. Correlating this with insect preference showed that predatory mites selected plants which were infested and water-stressed over infested but not water-stressed plants (Takabayashi et al 1994a).

The variation between plant species in the volatiles released with herbivore feeding and the generality of this response in the plant kingdom cannot yet be satisfactorily addressed because only a limited number of crop plants have been studied. Among crop plants that have been examined, parasitoid responses to leaf damage by different herbivores is variable.

The parasitoid *Microplitis croceipes*, which is strongly attracted to corn seedlings incubated in beet armyworm oral secretion or in buffer containing volicitin (the pure elicitor isolated from beet armyworm oral secretions), cannot parasitize beet armyworms. This and similar results with other parasitoids (Turlings et al 1993, 1999) suggested that a plant would release the same blend of volatiles regardless of the herbivore species feeding on it. However, in field trials, *Cardiochiles nigriceps* females clearly distinguished between tobacco plants damaged by their host *Heliothis virescens* and those infested by *Helicoverpa zea*, which is not a host (De Moraes et al 1998). This preference of the parasitoids for plants damaged by their hosts persisted even when all caterpillars and damaged leaves were removed from the plants, leaving only the undamaged leaves from which volatiles are released systemically. Also, the wasps were able to distinguish between cotton plants damaged by their hosts and those damaged by non-hosts. Furthermore, GC–MS analysis of volatiles released by plants damaged by the two herbivore species clearly showed differences in proportions of compounds in blends induced by *H. virescens* and those induced by *H. zea* feeding on both cotton and tobacco. Thus plants respond differently to damage by different species of herbivores and release signals that can be distinguished by specialist parasitoids. This strongly suggests that different herbivore species produce different elicitors.

Elicitors of plant volatiles from insect herbivores

Thus far only two elicitors of plant volatiles have been identified in the oral secretions of insect herbivores. Mattiacci et al (1995) reported that a β-glucosidase in the saliva of *Pieris brassicae* caterpillars elicits the release of volatiles from cabbage leaves. More recently Alborn et al (1997) identified volicitin (N-[17-hydroxylinolenoyl]-L-glutamine) (Fig. 3) as the major active elicitor in the oral secretion of beet armyworm larvae. Synthesized and natural volicitin induce corn (*Zea mays* L.) seedlings to release the same blend of volatile terpenoids and indole as released when they are damaged by caterpillar feeding (Alborn et al 1997). This blend of volatile compounds attracts parasitic wasp females that attack the caterpillars.

It has been suggested that jasmonic acid, which is produced from linolenic acid by the octadecanoid signalling pathway, may be a key regulatory component in the transduction sequence that triggers synthesis and release of volatile compounds by plants (Krumm et al 1995). Jasmonates also stimulate other physiological and defensive processes in plants (Farmer & Ryan 1992a,b, Krumm et al 1995), and the amino acid conjugates of jasmonic acid are involved in physiological and developmental processes in many plants (Krumm et al 1995, Kramell et al 1995). Therefore, the presence of an elicitor which is an octadecatrienoate conjugated to an amino acid suggests that the elicitor molecule interacts with the octadecanoid pathway in the herbivore-damaged plants.

Although volicitin has not been found in plants, we have recently demonstrated that the caterpillar synthesizes it by adding a hydroxyl group and glutamine to linolenic acid obtained directly from the plant on which the caterpillar feeds (Paré et al 1998). This strongly suggests that these molecules play an important, but still unknown, role either in the metabolism or some other process critical to the life of the herbivorous insects. It is also interesting to note that the plant is providing the linolenic acid, which is essential for most lepidopteran larvae (Stanley-Samuelson 1994, and references therein), with which the insect makes an elicitor of plant chemical defences, seemingly detrimental to the insect. The full implications of this are not yet understood.

The use of volatile plant compounds as chemical cues by natural enemies of herbivores to locate their hosts has been identified in a wide range of agricultural species. Although two elicitors have been identified from caterpillars that feed on plants, little is known about the initial signalling pathway (or pathways) that lead

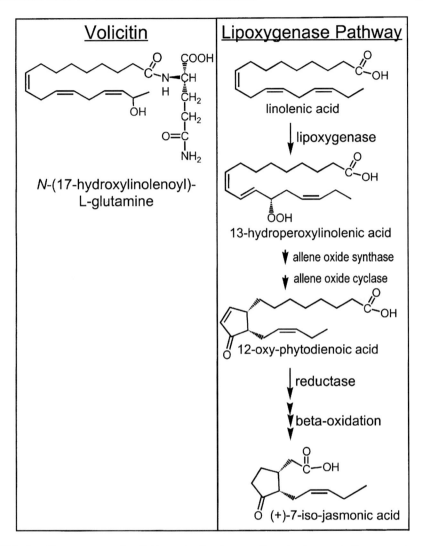

FIG. 3.   The structure of volicitin and select intermediates in the metabolic conversion of linolenic acid to jasmonic acid.

to the synthesis and emissions of these volatile compounds. What role the octadecanoic-based elicitor volicitin may have in regulating metabolism or gene action of plant-derived lipid signals is not clear. How hydrolytic enzymes such as glucosidases trigger the release of volatile metabolites is also unknown, especially since the volatiles released do not appear to be stored as sugar-linked precursors. Elucidation of the transduction pathway(s) and mechanisms by which the synthesis and release of volatiles are induced in plants in response to herbivore damage will be a fertile area for research for several years.

Knowledge gained from this research should be useful in designing new ecologically sound approaches to insect pest management. Clearly, crop varieties that release larger amounts of volatiles when attacked by insect herbivores, and thus become easier for parasitoids and predators to find, can be developed through plant breeding or genetic engineering. When used with other ecologically sound farming practices, such as conservation tillage, that provide reservoirs of beneficial insects, such varieties should enhance the effectiveness of biological control. Also, it may eventually be possible to evoke a response by the plants at a time of our choosing to make them more resistant to a sudden increase in an insect population. Finally, a better understanding of the interactions in the tritrophic systems may allow us to choose better plant species for intercropping, and/or more effective parasitoids and predators for biological control. Also, as we progress in our understanding of these systems other approaches may be discovered that are not evident now.

References

Alborn HT, Turlings TCJ, Jones TH, Stenhagen G, Loughrin JH, Tumlinson JH 1997 An elicitor of plant volatiles from beet armyworm oral secretion. Science 276:945–949

Boland W, Feng Z, Donath J, Gabler A 1992 Are acyclic C_{11} and C_{16} homoterpenes plant volatiles indicating herbivory? Naturwissenswchaften 79:368–371

De Moraes CM, Lewis WJ, Paré PW, Alborn HT, Tumlinson JH 1998 Herbivore-infested plants selectively attract parasitoids. Nature 393:570–573

Dicke M 1994 Local and systemic production of volatile herbivore-induced terpenoids: their role in plant–carnivore mutualism. J Plant Physiol 143:465–472

Dicke M, van Deek TA, Posthumus MA, Ben Dom N, Van Dokhoven H, De Groot AE 1990 Isolation and identification of volatile kairomone that affects acarine predator–prey interactions. Involvement of host plant in its production. J Chem Ecol 16:381–396

Farmer EE, Ryan CA 1992a Octadecanoid precursors of jasmonic acid activate the synthesis of wound-inducible proteinase inhibitors. Plant Cell 4:129–134

Farmer EE, Ryan CA 1992b Octadecanoid derived signals in plants. Trends Cell Biol 2:236–241

Kramell R, Atzorn R, Schneider G et al 1995 Occurrence and identification of jasmonic acid and its amino conjugates induced by osmotic stress in barley leaf tissue. J Plant Growth Regul 14:29–36

Krumm T, Bandemer K, Boland W 1995 Induction of volatile biosynthesis in the lima bean (*Phaseolus lunatus*) by leucine- and isoleucine conjugates of 1-oxo- and 1- hydroxyindan-4-carboxylic acid: evidence for amino acid conjugates of jasmonic acid as intermediates in the octadecanoid signaling pathway. FEBS Lett 377:523–529

Lewis WJ, van Lenteren JC, Phatak SC, Tumlinson JH 1997 A total system approach to sustainable pest management. Proc Natl Acad Sci USA 94:12243–12248

Loughrin JH, Manukian A, Heath RR, Turlings TCJ, Tumlinson JH 1994 Diurnal cycle of emission of induced volatile terpenoids by herbivore-injured cotton plants. Proc Natl Acad Sci USA 91:11836–11840

Loughrin JH, Manukian A, Heath RR, Tumlinson JH 1995 Volatiles emitted by different cotton varieties damaged by feeding beet armyworm larvae. J Chem Ecol 21:1217–1227

Mattiacci L, Dicke M, Posthumus MA 1995 β-glucosidase: an elicitor of herbivore-induced plant odor that attracts host-searching parasitic wasps. Proc Natl Acad Sci USA 92:2036–2040

Paré PW, Tumlinson JH 1997a De novo biosynthesis of volatiles induced by insect herbivory in cotton plants. Plant Physiol 114:1161–1167

Paré PW, Tumlinson JH 1997b Induced synthesis of plant volatiles. Nature 385:30–31

Paré PW, Tumlinson JH 1998 Cotton volatiles synthesized and released distal to the site of insect damage. Phytochemistry 47:521–526

Paré PW, Alborn HT, Tumlinson JH 1998 Concerted biosynthesis of an insect elicitor of plant volatiles. Proc Natl Acad Sci USA 95:13971–13975

Röse USR, Manukian A, Heath RR, Tumlinson JH 1996 Volatile semiochemicals released from undamaged cotton leaves: a systemic response of living plants to caterpillar damage. Plant Physiol 111:487–495

Stanley-Samuelson DW 1994 Prostaglandins and related eicosanoids in insects. Adv Insect Physiol 24:115–212

Takabayashi J, Dicke M, Posthumus MA 1991 Variation in composition of predator attracting allelochemicals emitted by herbivore-infested plants; relative influence of plant and herbivore. Chemoecology 2:1–6

Takabayashi J, Dicke M, Takahashi S, Posthumus MA, Van Beek TA 1994a Leaf age affects composition of herbivore-induced synomones and attraction of predatory mites. J Chem Ecol 20:373–386

Takabayashi J, Dicke M, Posthumus MA 1994b Volatile herbivore-induced terpenoids in plant-mite interactions: variation caused by biotic and abiotic factors. J Chem Ecol 20:1329–1354

Turlings TCJ, Tumlinson JH 1992 Systemic release of chemical signals by herbivore-injured corn. Proc Natl Acad Sci USA 89:8399–8402

Turlings TCJ, Tumlinson JH, Lewis WJ 1990 Exploitation of herbivore-induced plant odors by host-seeking parasitic wasps. Science 250:1251–1253

Turlings TCJ, Tumlinson JH, Heath RR, Proveaux AT, Doolittle RE 1991 Isolation and identification of allelochemicals that attract the larval parasitoid, Cotesia marginiventris (Cresson), to the microhabitat of one of its hosts. J Chem Ecol 17:2235–2251

Turlings TCJ, McCall PJ, Alborn HT, Tumlinson JH 1993 An elicitor in caterpillar oral secretions that induces corn seedlings to emit chemical signals attractive to parasitic wasps. J Chem Ecol 19:411–425

Turlings TCJ, Loughrin JH, McCall PJ, Röse US, Lewis WJ, Tumlinson JH 1995 How caterpillar-damaged plants protect themselves by attracting parasitic wasps. Proc Natl Acad Sci USA 92:4169–4174

Turlings TCJ, Lengwiler UB, Bernasconi ML, Wechsler D 1998 Timing of induced volatile emissions in maize seedlings. Planta 207:146–152

Turlings TCJ, Alborn HT, Loughrin JH, Tumlinson JH 1999 Volicitin, an elicitor of maize volatiles in the oral secretion of Spodoptera exigua: I. Its isolation and bio-activity. J Chem Ecol, in press

DISCUSSION

Gierl: In some plant resistance reactions the plants have selected compounds of the pathogen that are required for full virulence. Does volicitin also have a function in the life cycle of the herbivore and was it therefore selected by the plant as a signal to fight against the insect?

Tumlinson: That is a very interesting question, and it is one we would like the answer to. It is tempting to think that volicitin has a critical role for the caterpillar. It is also interesting that the plant is sort of 'feeding' the caterpillar: linolenic acid is an essential fatty acid for Lepidopterans. They need it in their diet, and then they're making some compound that triggers the defensive reactions of the plant.

Lamb: I was wondering about the same issue. In the pathogen field it is becoming clear that avirulence genes seem to have primary functions as pathogenicity factors.

Haukioja: Another dimension in this 'spit ecology' is that the spit may also be important in communication between the larvae. It was found about 30 years ago by Gruys (1970) that when pine looper larvae were reared together they exchanged spit droplets, which caused them to grow differently from larva kept separate. We have repeated this work with another geometrid moth and found that keeping larvae together for a couple of days makes them behave for their whole lifetimes as crowded larvae. We are not absolutely sure whether it's the spit which triggers the crowding response, but this is the conclusion drawn by the Dutch group. This phenomenon is not species-specific—different species exchange spit—and after this happens growth is different than that of solitary individuals.

Tumlinson: It is certainly possible that compounds in the spit have some sort of intraspecific communication function. However, the Spodoptera are naturally gregarious, so this may not be relevant to them.

Boland: What do you know about the concentration dependency of volicitin and volatile induction? If you reduce the concentration of volicitin gradually, is there a gradual decrease in volatiles or are classes of volatiles progressively shut off at each stage?

Tumlinson: We do see a decrease in volatile production as we reduce the volicitin concentration, but we seem to get a sort of threshold response. We haven't really looked at this in a way that would answer your question about classes.

Städler: Do your *Heliothis* (*Helicoverpa*) species respond to linolenic acid?

Tumlinson: We haven't demonstrated that linolenic acid stimulates the larva to feed in any case.

Städler: Have you tried?

Tumlinson: No, we have not done that. We know from the literature that linolenic acid is an essential fatty acid for Lepidopteran larvae, but I don't think it has ever been shown to be a feeding stimulant.

Städler: We showed long ago that in *Manduca sexta*, if you raise caterpillars on artificial diets the linolenic acid can have an inducing effect (Städler & Hanson 1978).

Scheel: Do you know how many different plants are able to perceive volicitin? Mammals and plants normally hydroxylize lipophilic compounds or conjugate them to amino acids and sugars, in order to handle them. Is it possible that the herbivore just uses that mechanism to handle that lipophilic compound?

Tumlinson: That is a good possibility. Certainly, putting an amino acid onto this fatty acid makes it much more soluble in aqueous systems, and probably makes it easier to handle. We obviously don't know whether or not that's the function of volicitin.

We haven't tested the purified linolenic acid on many plants. You can put feeding beet armyworms on corn and cotton and the plants respond to the feeding caterpillars. We haven't had a chance to test purified volicitin on a whole series of plant species, but it is something we would like to do.

Lamb: Is this diurnal rhythm of the production of the induced volatiles just to do with the feeding behaviour of the caterpillar?

Tumlinson: You can take the caterpillar off after the first day and you will continue to get this diurnal rhythm for a couple of days, although it decreases on subsequent days.

Dicke: You have shown that there is this diurnal rhythm with chemical data. Throughout the day the composition of the blend changes, so the information that the parasitoids receive changes throughout the day. But what evidence is there that this affects the behaviour of the parasitoids? Under natural circumstances the parasitoids are not active during the night. However, the amounts that are released during the night will still be enough for the parasitoids to respond to them, because if the GC will detect them the parasitoids will. What do you know about the response of parasitoids throughout the day?

Tumlinson: We don't have any data that suggest that the different blends trigger different responses in the parasitoids at different times of the day.

Dicke: This might be very important in relation to your work with different herbivores on the same plant. There you show blends that differ in composition as well as in total amount have different effects on the parasitoids. You showed behavioural data for *Cardiochiles nigriceps* that has preference for *H. virescens*. Would *Microplitis* take the opposite preference towards odours released by plants infested by *H. zea*?

Tumlinson: *Microplitis* attacks both species; it is not highly specialized. So it probably wouldn't give you the answer you're looking for.

Dicke: Wouldn't you be able to prime it to one or the other through learning and then see whether you could train them the other way?

Tumlinson: We probably could.

Poppy: It would be interesting to see whether the observations in the field fit in with this.

Gershenzon: Is there a diurnal rhythm to parasitoid activity and, if so, what is the timing of that rhythm with respect to induced volatile production?

Tumlinson: The literature said that the parasitoids are active during the day, primarily late morning and early afternoon.

Gershenzon: Interestingly, you and Paul Paré have also shown that there is a tight coupling between CO_2 fixation and monoterpene production (Paré & Tumlinson 1997). It seems as if a physiological constraint may be operating here: no photosynthesis means no monoterpenes. These compounds may have to be produced when photosynthesis is operating, which in turn is when parasitoids active.

Tumlinson: That may just be an artefact of how we set up our timing intervals. We were doing 3 h intervals, but there is a lot to be said for using shorter intervals. The other thing is, the more light there is the more volatiles are produced. If you bring these plants into the laboratory you don't get high levels of light, but if you put them out in the greenhouse on a bright sunny day high levels of volatiles are produced. Light is critical to the synthesis of these compounds.

Turlings: In Florida, where it gets pretty hot, the wasps will search during the day but mostly in the morning or late afternoon, so this is relatively nicely tied in with the diurnal rhythm of the production of the volatiles. During the middle of the day when the highest amounts of volatiles are released, the wasps have usually stopped searching because of the heat. I don't think we should think too much about this diurnal rhythm of volatile production being tied in with the wasps. It most likely has to do with photosynthesis.

Poppy: Aphid parasitoids in a wind tunnel are most reactive to plant volatiles 4–7 h into the photophase.

Bohlmann: How much of the diurnal rhythm is actually a temperature effect?

Turlings: That's a good point. We haven't done that control yet.

Baldwin: We have done that experiment with *Nicotiana attenuata*, and temperature is not responsible for the diurnal rhythms.

Poppy: Jim Tumlinson, what is the composition of the fall armyworm spit? How is it causing the changes in ratios of the induced plant volatiles? I'm confused about how it causes changes in ratios of the induced cotton volatiles when it actually has volicitin, just as the case for the beet armyworm.

Tumlinson: We really don't know. We haven't analysed and identified the compounds in fall armyworm spit. Obviously, we've done HPLC and we see similar type peaks, but we don't know yet what effects different components of the spit have on production of different volatile compounds. We've tried to look at just proportions of compounds in the spit of the beet armyworm to see what happens there. We don't have all these data in yet, and there are some problems in that we don't find consistent proportions of all the different components in the spit of the beet armyworm. As yet, we really don't understand how these compounds interact with the plant and trigger these different biosynthetic processes.

References

Gruys P 1970 Growth of *Bupalus piniarius* (Lepidoptera, Geometridae) in relation to larval population density. Centre for Agricultural Publishing and Documentation, Wageningen, The Netherlands

Paré PW, Tumlinson JH 1997 *De novo* biosynthesis of volatiles induced by insect herbivory in cotton plants. Plant Physiol 114:1161–1167

Städler E, Hanson FE 1978 Food discrimination and induction of preference for artificial diets in the tobacco hornworm, *Manduca sexta*. Physiol Entomol 3:121–133

Induced biosynthesis of insect
semiochemicals in plants

Wilhelm Boland, Thomas Koch, Thomas Krumm, Jörn Piel and Andreas Jux

Max Planck Institut für Chemische Ökologie, Tatzendpromenade 1a, D-07745 Jena, Germany

Abstract. Plants under attack by a herbivore may emit characteristic volatiles that are implicated in the attraction of the natural enemies of the herbivore. The signal cascade between leaf damage and the volatile production is stimulated by high- or low-molecular-weight elicitors from the secretions of the herbivore. Besides compounds from the octadecanoid signalling pathway, several structurally non-related amino acid conjugates such as the bacterial phytotoxin coronatine, the synthetic indanoyl-isoleucine, or amino acid conjugates of linolenic acid likewise induce volatile biosynthesis. Minor changes in the amino acid moiety may result in different volatile profiles (sesqui- and diterpenoids), attributing to the amino acid substructure a specific role for the recognition and the selective induction. Volatile terpenoids (mono- and diterpenoids) are synthesised *de novo* along the novel deoxy-D-xylulose (DOX) pathway, while the biosynthesis of sesquiterpenes may be fuelled from both the DOX- and the mevalonate pathway. This finding may be of importance for the plant defence in case of introduction of inhibitors together with the salivary secretion of herbivores into the leaf tissue.

1999 Insect–plant interactions and induced plant defence. Wiley, Chichester (Novartis Foundation Symposium 223) p 110–131

Plants utilize a variety of defensive strategies ranging from mechanical to chemical weapons, some of which are effective obstacles to herbivores and pathogens. While the mechanical devices generally represent constitutive traits, the chemical protection is not necessarily constitutive, but may be produced on demand after infestation or herbivore damage. Moreover, plants are sessile organisms and this property in particular may have forced the evolution of the multitude of adaptation mechanisms to (environmental) stresses we see today. To overcome the peculiar constraints that result from their stationary way of life, plants utilize volatiles for long-distance interactions. For example, volatiles serve as attractants for pollination (Langenheim 1994, Harrewijn et al 1995, Knudsen et al 1993) and seed dispersal (Midgely & Bond 1991) or they attract prey to carnivorous plants (Kite 1995). Herbivore-induced plant volatiles can even serve as cues to direct predators into the vicinity of their prey (Dicke et al 1990, Turlings et al 1990);

this signalling of the plant to higher trophic levels has been interpreted as 'a plant's cry for help' (Dicke & Sabelis 1992). Emission of volatiles, such as ethylene, methyl jasmonate or methyl salicylate by plants under attack may even induce defence and resistance genes in undamaged neighbouring plants (Farmer & Ryan 1990, Miksch & Boland 1996, Shulaev et al 1997). Besides (info)chemicals in the gas phase addressing insects or defending against insects over distance, other generally less volatile compounds are deposited on the leaf surface and serve there as another barrier against attacking pathogens and herbivores prior to the first serious damage of the leaf tissue (Stanjek et al 1998 and references cited therein). The permanent adaptation of a plant to the actual environmental conditions or stresses necessitates extensive reprogramming of gene expression. This, however, first requires the ability to perceive external events like, for example, infestation, herbivory, drought, salinity or temperature extremes. Next, the organism must be able to transduce the experience of the external event into an inter- and/or intracellular signal and finally into locally or systemically altered gene expression.

The octadecanoid signalling pathway

Among several signals currently known to be involved in plant stress responses, the octadecanoid signalling cascade is of special importance (Sembdner & Parthier 1993, Yoshihara & Greulich 1999). The cascade is linked to the external events in a still unknown fashion and starts with the formation of $(13S)$-hydroperoxylinolenic acid from free and/or bound linolenic acid (Vick & Zimmermann 1984, Feussner & Kindl 1994, Feussner et al 1995). Next, the hydroperoxide is converted into an unstable allene oxide which is rapidly converted in phytodienoic acid (PDA) (Hamberg 1988), recently recognized as the most important signal for plant mechanoreceptors (Falkenstein et al 1991, Weiler 1997). PDA is reduced and then degraded by three consecutive β-oxidation cycles to (epi)jasmonic acid (Fig. 1). According to Fig. 1 the lipid-based signalling pathway comprises at least three structurally different types of signalling compounds: (a) acyclic fatty acids, (b) cyclopentanoid C_{18} fatty acids, and (c) the cyclopentanoid C_{12} fatty acid (epi)jasmonic acid.

Moreover, recently amino acid conjugates of linolenic acid, e.g. volicitin (Alborn et al 1997) and of jasmonic acid (JA) (Kramell et al 1995, Tamogami et al 1997, Krumm 1998) were identified as another group of biologically active compounds belonging to this pathway. At least in principle, each of the different intermediates of the pathway could serve as a discrete signal for activation of independent metabolic activities. An example of a differential use of intermediates from the pathway is provided by the volatile induction in the lima bean (*Phaseolus lunatus*) (Krumm 1998). Thus, treatment of freshly cut plantlets

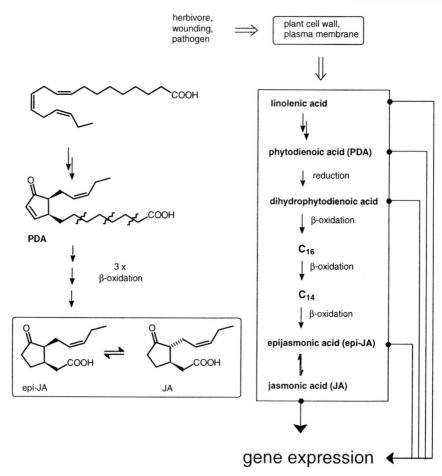

FIG. 1. Octadecanoid signalling pathway in plants. The molecular interfaces between the primary leaf damage and the octadecanoid pathway and from there to gene expression are not known. Some fatty acid derivatives also occur as conjugates with amino acids.

with solutions of JA resulted in the emission of the complex pattern of volatiles shown in Fig. 2A (Hopke et al 1994, Boland et al 1995), but application of an early precursor of the pathway like, for example, free linolenic acid stimulated only the biosynthesis of the C_{11} homoterpene 4,8-dimetyl-nona-1,3,7-triene (DMNT) and, much more pronounced that of the C_{16} compound 4,8,12-trimethyltrideca-1,3,7,11-tetraene (TMTT) (Fig. 2B). The biosynthesis of the latter is clearly not induced by JA. If lipid peroxidation is inhibited by pre-treatment with phenidone (Peña-Cortés et al 1993, Doares et al 1995) prior to addition of linolenic acid, no induction of volatile biosynthesis is observed, thus linking the induction of homoterpene biosynthesis to compounds downstream of linolenic acid. Next, if PDA, the first cyclopentanoid member of the pathway was applied, the two homoterpenes were produced together with several other components of the volatile profile shown in Fig. 2A. Thus, it is reasonable to assume, that PDA and probably another signal between PDA and linolenic acid induce the diterpenoid metabolism leading to the C_{16} homoterpene (Fig. 2B), while jasmonate is the master switch for most of the other compounds. Compounds downstream of PDA, e.g. dihydro-PDA and the subsequent β-oxidation products (Fig. 1), were much less active or inactive (Hopke 1997). At least in the lima bean, JA, the last member of the octadecanoid family, appears to be the most powerful signal with respect to volatile induction (Hopke et al 1994, Boland et al 1995).

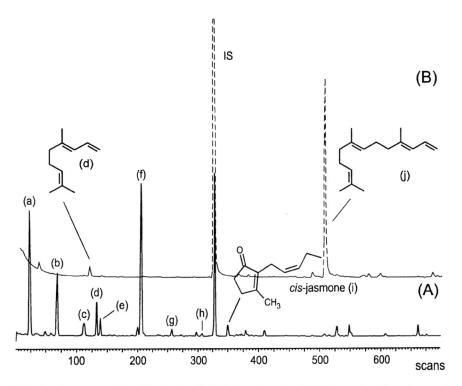

FIG. 2.   Jasmonate (A) and linolenic acid (B) induced profile of volatiles emitted from leaves of *P. lunatus* (Hopke et al 1994, Boland et al 1995). Identification of compounds: (a) (3Z)-hexenyl acetate; (b) β-ocimene; (c) linalool; (d) 4,8-dimetyl-nona-1,3,7-triene (DMNT); (e) $C_{10}H_{14}$; (f) $C_{10}H_{16}O$; (g) indole; (h) (3Z)-hexenyl methylbutanoate; IS, internal standard (1-bromodecane); (i) *cis*-jasmone; (j) 4,8,12-trimethyltrideca-1,3,7,11-tetraene (TMTT). Volatiles were collected by absorption onto carbon traps (Donath & Boland 1995) or by inserting an SPME fibre (Zhang et al 1994) into the gas phase. Separation and identification of the compounds was achieved by GC-MS.

Other plants, like for example *Zea mays*, did not show such a differential induction of volatile patterns, and, hence, the extent and significance of the differential induction in plants still remains to be established. Unlike the apparently plant-specific differential induction, JA appears to be active in a multitude of plants ranging from archetype (ferns, gingko) to modern species (Boland et al 1995).

Coronatine and indanones as molecular tools for the elicitation of defensive pathways

In addition to compounds from the octadecanoid family, the bacterial phytotoxin coronatine (Ichihara et al 1977, Nüske & Bublitz 1993), first isolated by Ichihara and co-workers, also elicits volatile production in plants at exceptionally low concentrations (Boland et al 1995). The phytotoxin is produced by pathovars of *Pseudomonas syringae* (e.g. *tomato, glycinea, atropurpurea*) (Nüske & Bublitz 1993). The structurally much simpler and accordingly designed synthetic analogue 1-oxo-indanoyl-isoleucine (IN-Ile) is also a powerful elicitor of volatile production and can be generally used instead of the more difficult to assess coronatine (Fig. 3; Krumm et al 1995). The major advantage of the amino acid conjugates of IN-Ile is the *complete lack of bioactivity of their building blocks*, namely the amino acid and the 1-oxo-indan-4-carboxylic acid.

Therefore, activity can be attributed to the conjugates, rather than to their hydrolysis products. Instead of the free acid IN-Ile, the methyl or allyl esters also showed excellent inducing properties. The high activity of the non-polar esters might be attributed (i) to their enhanced membrane permeability and (ii) to the fact that they represent reactive esters which are readily saponified after a membrane passage. Interestingly, only the conjugates containing the *L*-amino

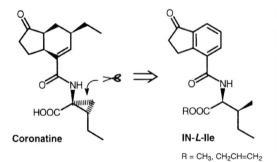

amino acid	induction of volatiles
L-Ile	+++
D-Ile	-
L-allo-Ile	+
L-Leu	++
D-Leu	-

R = CH₃, CH₂CH=CH₂

FIG. 3.   1-Oxo-indanoyl-isoleucine (IN-Ile) and other amino acid conjugates as mimics of coronatine. Biological activity was determined by the induction of volatile biosynthesis.

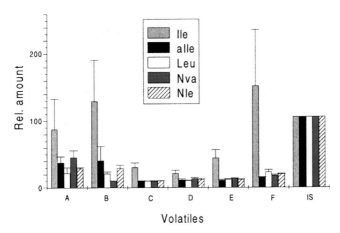

FIG. 4. Relative amounts of emitted volatiles after induction with various amino acid conjugates of 1-oxo-indan-4-carboxylic acid. (A) (3Z)-hexenyl acetate, (B) β-ocimene, (C) linalool, (D) DMNT, (E) $C_{10}H_{14}$, (F) $C_{10}H_{16}O$, (IS) internal standard (1-bromodecan, =100).

acids Ile and Leu were active (Fig. 4). Others, containing, for example, the amino acids Val, Gln, Phe, Met or norvaline (Nva) and norleucine (Nle) did not induce volatile biosynthesis.

The conjugate with isoleucine (IN-Ile) induced a volatile pattern matching that of a JA treatment, but the conjugate with coronamic acid (IN-Cor), in addition, induced the biosynthesis of the C_{16} homoterpene. Systematic studies with other amino acid conjugates revealed that several different signalling circuits (*vide supra*) must exist in the defence responses of the lima bean, since either volatile patterns resembling those from JA, PDA, or linolenic acid treatment could be provoked by changing the amino acid moieties (Krumm 1998).

Although the amino acid conjugates coronatine and IN-Ile appear to be structurally related to the amino acid conjugates of JA (Kramell et al 1995, Tamogami et al 1997), such as JA-Ile, they cannot be considered as analogues of the corresponding JA conjugates (Fig. 5). While the non-hydrolysable amides IN-Cor and IN-(α-Me)Leu strongly provoke the biosynthesis of volatiles, the related conjugates with jasmonic acid JA-Cor and JA-(α-Me)Leu failed to induce biosynthetic activities (Krumm 1998). Another important structural aspect for the biological activity of IN-Ile is the need for the presence of the amide bond. Synthesis of the deoxy-carba analogue of IN-Ile provided an inactive compound. The non-hydrolysable deoxy-carba JA-Ile was also inactive with respect to volatile induction in the lima bean, but the compound provoked an up-regulation of some house-keeping genes such as those coding for rubisco and chlorophyll-a/b-binding proteins when applied to barley leaves (Wasternack et al 1998).

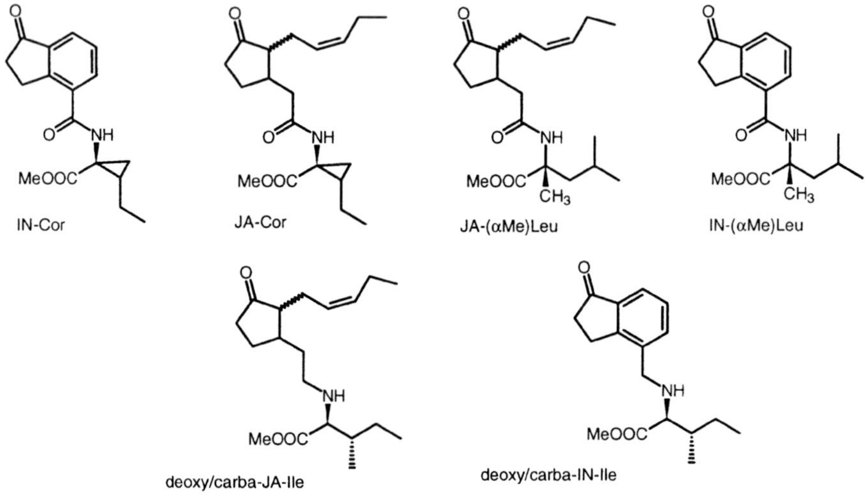

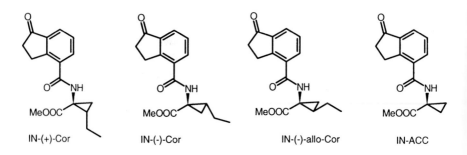

FIG. 5.   Upper row: non-hydrolysable amino acid conjugates of 1-oxo-indan-4-carboxylic and jasmonic acid. Lower row: deoxa-carba analogues of JA-Ile and IN-Ile.

A highly specific interaction of IN-(+)-Cor with macromolecular receptors is independently supported by the observation that only the conjugate with the natural (+)-coronamic acid was active (Fig. 6). All other stereoisomers, e.g. IN-(−)-Cor or IN-(−)-*allo*-Cor and a conjugate with aminocyclopropyl amino acid (IN-ACC), lacking the ethyl substituent on the cyclopropane moiety, were completely inactive (Hopke 1997).

To test for the possibility that the conjugates induce the octadecanoid pathway, lima beans were pre-treated with coronatine or IN-Ile followed by analysis of the

FIG. 6.   Conjugates of 1-oxo-indan-4-carboxylic acid with diastereomers of coronamic acid and methyl aminocyclopropanoate.

internal JA level of the plants (Baldwin et al 1997). A typical JA profile representing the coronatine or IN-Ile treatment is shown in Fig. 7.

Despite of a minor increase of the internal JA level after the onset of the experiment, (mechanical stress of cutting the plantlet), no increase of the internal JA level could be observed after application of the conjugate. For comparison, treatment of lima bean leaves through the petiole with the high molecular elicitor cellulysin transiently enhanced the JA level to *c*. 350–400 ng/g fresh weight (Piel et al 1997). A significant increase of the internal JA level was also reported to occur after herbivore feeding on tobacco leaves (McCloud & Baldwin 1997). Considering these data it becomes obvious that coronatine and IN-Ile may bind to specific macromolecular targets triggering activities related to that of PDA and JA, but it remains to be established whether or not these targets are different from those present for recognition of the natural octadecanoid signals.

Biosynthesis of induced terpenoid volatiles: homoterpenes

A considerable amount of the volatiles released from damaged and/or infected plants are terpenoids. In higher plants, at least two subcellular compartments are known that produce isoprenoids (Gershenzon & Croteau 1993, Chappell 1995). Enzymes in the cytoplasm synthesize sesquiterpenes, sterols and triterpenoids, and enzymes of the plastidic compartment produce isoprene, monoterpenes, diterpenes and carotenoids. It is generally accepted that at least the final biosynthetic steps are bound to the site of the occurrence of the compounds. Although all prenyl lipids are synthesized from isopentenyl diphosphate (IPP)

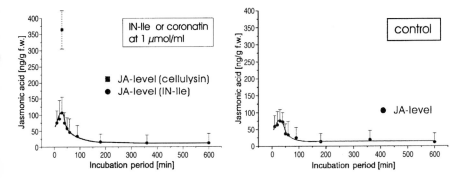

FIG. 7. Analysis of the internal JA level of cut lima bean plantlets after pre-treatment with solutions of the conjugates through the petiole. Left: (●) internal JA level after treatment with IN-Ile or coronatine, (■) internal JA-level after treatment with cellulysin (Piel et al 1997). Right: JA level (●) of an untreated control plant in tap water. The transient increase of the JA level after the onset of the experiment is due to the mechanical stress of cutting (cf. control).

and dimethylallyl diphosphate (DMAPP) as the universal building blocks, work of recent years has shown that the early steps towards IPP are entirely different in the cytosol and the plastids (Rohmer et al 1993, Arigoni et al 1997, Lichtenthaler 1998). As outlined in Fig. 8, in the cytoplasm the long-known mevalonate-dependent pathway is operative, while in plastids IPP is generated in close association with the catabolism of glucose by condensation of glyceraldehyde-3-phosphate (GA-3P) with activated pyruvate to yield 1-deoxy-D-xylulose-5-phosphate (Arigoni et al 1997). Following a skeletal rearrangement and several elimination steps IPP is obtained as the central intermediate (Giner et al 1998).

Up to now, most of the results summarized in Fig. 8 were obtained from inhibitor and feeding studies using [^{13}C]- or [^{14}C]-glucose or labelled deoxy-D-xylulose (DOX). Since the first genes of enzymes from the novel mevalonate-independent pathway have been identified and cloned, rapid progress and

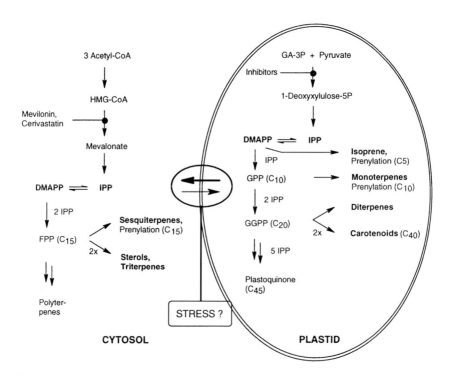

FIG. 8. Localization of the mevalonate-dependent and the mevalonate-independent biosynthesis of terpenoids in cytosolic and plastidic compartments of the cell (modified after Lichtenthaler et al 1997). The extent of precursor exchange between the compartments is not known.

consolidation of the localization of these enzymes is to be expected in near future (Lange et al 1998).

According to Fig. 8 the herbivore- and/or pathogen induced *de novo* biosynthesis of terpenoids should be, thus, either fuelled from the mevalonate or from DOX pathway. Accordingly, the two homoterpenes DMNT and TMTT (Fig. 2A/B) representing typical and widespread herbivore-induced terpenoids, should come from different pools, since DMNT is of sesquiterpenoid origin while TMTT is a degradation product of the diterpenoid geranyllinalool (Fig. 9; Gäbler et al 1991, Boland et al 1998a). The enzymes required for the oxidative degradation of the two terpenoid alcohols to the homoterpenes are apparently constitutively present in many plants (Boland et al 1992), since the administration of labelled precursors without exception resulted in an almost instantaneous production of the hydrocarbons.

On the other hand, feeding of early precursors such as mevalonic acid (MVA) or DOX did not result in the production of homoterpenes unless an elicitor like JA was applied or herbivore feeding was allowed. To determine the actual origin and site of biosynthesis of DMNT and TMTT (cf. Figs 8 and 9), differently labelled C_5 precursors, $[D_5]$-mevalolactone ($[D_5]$-MVA]) (Cane et al 1990) and $[D_2]$-deoxy-D-xylulose ($[D_2]$-DOX) (Piel & Boland 1997) were synthesized and administered to the plant followed by elicitation with JA. Dependent on the engaged pathway and type of utilized precursor, the *de novo* synthesized homoterpenes DMNT and TMTT were expected to comprise different degrees of labelling as outlined in Fig. 10.

The positions and number of the incorporated deuterium atoms were attributed following the general outline of terpenoid biosynthesis utilizing IPP and DMAPP as the isoprenoid building blocks. As a matter of fact, after pre-incubation with $[D_2]$-DOX generally an exceedingly high degree of labelling was observed in all emitted terpenoids (Boland et al 1998b, Piel et al 1998). As demonstrated in Fig. 11a,b the origin of the metabolites could be reliably determined by mass spectrometry owing to their different degree and pattern of isotopic labelling.

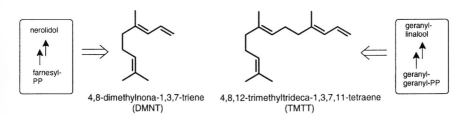

FIG. 9.   Biosynthetic origin of the two homoterpenes DMNT and TMTT from nerolidol and geranyllinalool.

For example, following administration of [D$_2$]-DOX to leaves of *P. lunatus* in conjunction with spider mite damage, [D$_3$]-DMNT, labelled to about 35% was obtained (Fig. 11a); [D$_5$]-TMTT from the same experiment showed a much higher degree of labelling (90%) (Piel et al 1998).

On the contrary, after pre-incubation with [D$_5$]-MVA and spider mite damage highly labelled (c. 82%) [D$_8$]-DMNT was formed (Fig. 11b), while this time the C$_{16}$ homoterpene failed to exhibit a high degree of labelling (c. 10%). In general, the incorporation of labelled MVA into the DMNT proved to be much less reproducible than the corresponding experiments with [D$_2$]-DOX. As demonstrated by the mass spectra of [D$_3$]- and [D$_8$]-DMNT resulting from the different precursors, the position and number of the isotopes within the metabolites were in complete agreement with pathways outlined in Fig. 10. The degree of labelling of the monoterpenoids (cf. Fig. 12) such as ocimene and linalool revealed a simpler pattern. Pre-incubation with [D$_2$]-DOX always resulted in highly labelled [D$_4$]-ocimene (90%) and [D$_4$]-linalool (90%), but pre-treatment

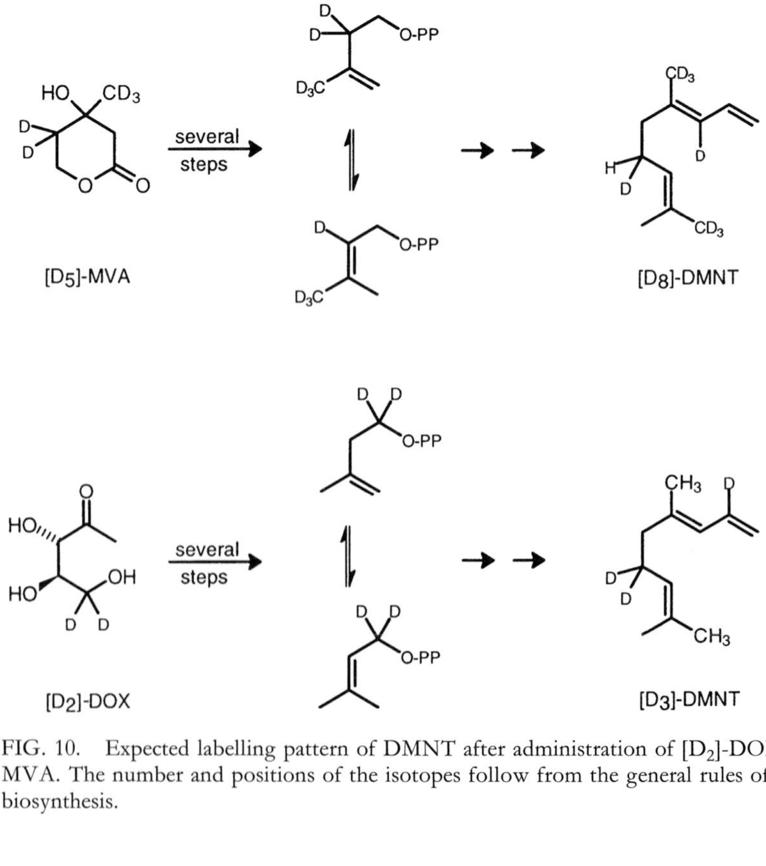

FIG. 10.   Expected labelling pattern of DMNT after administration of [D$_2$]-DOX or [D$_5$]-MVA. The number and positions of the isotopes follow from the general rules of terpenoid biosynthesis.

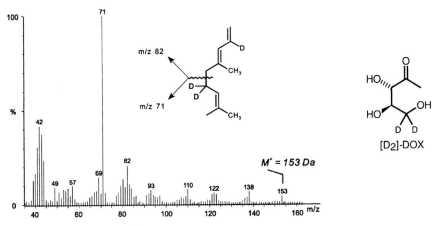

FIG. 11a. Mass spectrum of [D$_3$]-DMNT resulting from treatment of a lima bean leaf with [D$_2$]-DOX and subsequent damage by spider mite infestation.

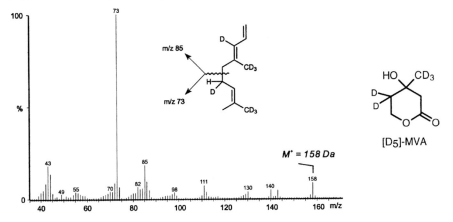

FIG. 11b. Mass spectrum of [D$_8$]-DMNT resulting from treatment of a lima bean leaf with [D$_5$]-MVA and subsequent damage by spider mite infestation or JA elicitation.

with [D$_5$]-MVA, if successful at all, furnished [D$_8$]-ocimene and [D$_8$]-linalool showing a low degree of labelling (less than 10%). The diterpenoid-derived TMTT followed the same trend, provided that the elicitation was achieved by spider mite damage; JA treatment failed to induce the biosynthesis of TMTT (*vide supra*). If the mevalonate-dependent pathway was blocked by application of an HMG-CoA reductase inhibitor like mevilonin or cerivastatin (Bischoff et al 1997) followed by administration of [D$_2$]-DOX, the degree of labelling of the emitted [D$_3$]-DMNT was significantly enhanced, but the extent of labelling of ocimene and linalool was not further increased (cf. Fig. 12) confirming the rather low importance of the MVA-pathway for the biosynthesis of monoterpenes (cf. Fig. 8).

The simultaneous use of both pathways, mevalonate-dependent and mevalonate-independent, *en route* to the alarm code DMNT may indicate a significant exchange of precursors (e.g. of isopentenyl-pyrophosphate) between the cytosolic and the plastidic compartments. This exchange could be of considerable importance for stressed plants to use increasing amounts of plastidic precursors to secure the cytosolic biosynthesis of sesquiterpene phytoalexins, if the MVA route is blocked or insufficient (cf. Fig. 8). Whether this flexible management of precursor pools is of general importance to the plant terpene biosynthesis, or whether a DOX-route also exists in the cytosol, remains to be explored.

The above findings were not limited to the dicotyledonous lima bean. High levels of incorporation of $[D_2]$-DOX after JA induction were also observed for tobacco (*Nicotiana plumbaginifolia*) and the monocotyledonous maize (*Zea mays*). The latter showed a high degree of labelling for both homoterpenes ($[D_3]$-DMNT, 80% and $[D_5]$-TMTT, 96%) when stimulated with JA. Even in the evolutionarily old gingko tree (*Gingko biloba*) the biosynthesis of JA-inducible sesquiterpenes such as linalool, copaene, caryophyllene and farnesene was predominantly achieved utilizing $[D_2]$-DOX as the precursor (Piel et al 1998).

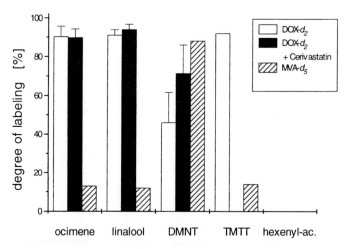

FIG. 12.   Degree of labelling of individual terpenoids after treatment of leaves of *P. lunatus* with $[D_2]$-DOX, $[D_5]$-MVA, and $[D_2]$-DOX in combination with the HMG-CoA reductase inhibitor cerivastatin (Bischoff et al 1997). Spider mite damage was used for induction of volatile biosynthesis. The data for ocimene, linalool and DMNT are also in accord with JA elicitation of volatile biosynthesis. The fatty acid metabolite hexenyl acetate is not labelled excluding randomization of the isotope label via degradative pathways (e.g. acetyl-CoA).

TABLE 1 Degree of labelling of constitutively emitted volatile terpenoids after treatment of freshly cut leaves or flowers with [D$_2$]-DOX

Eucalyptus globulus (leaf)		Clematis vitalba (flower)		Hedera helix (flower)	
Compound/ degree of labelling [%]		Compound/ degree of labelling [%]		Compound/ degree of labelling [%]	
1,8-cineole	0	β-pinene	30	Sabinene	25
Limonene	0	Limonene	24	β-myrcene	36
Ocimene	83	Ocimene	77	Limonene	20
DMNT	95	Oxoisophorone*	(0)	Oxoisophorone*	(0)
Caryophyllene	83	Dihydrooxo-isophorone*	(0)	Dihydrooxo-isophorone*	(0)
Nerolidol	94	Endo-borneol	15	Linalool	83
TMTT	99	Nerol	33	Linalooloxide	50
		α-terpineol	11	Farnesene	89
		α-verbenone	14		
		Caryophyllene	70		

Compounds marked with * lose their isotopic label through the later steps of oxidative degradation.

Comparable observations were made upon treatment of flowers or leaves, which constitutively emit volatiles, with [D$_2$]-DOX. Accordingly, volatile terpenoids from [D$_2$]-DOX-treated flowers of *Passiflora caerulea*, *Clematis vitalba*, *Hedera helix*, *Callicarpa japonica* and from leaves of *Eucalyptus globulus* showed a remarkable, but individually varying degree of labelling for most terpenoids (cf. Table 1).

Depending on their different individual metabolic history (storage and release from vacuoles, release from conjugates, mevalonate-dependent biosynthesis) the extent of labelling of the individual compounds varied considerably (0–100%).

The results presented here clearly demonstrate that the recently discovered mevalonate-independent pathway (DOX route or *Rohmer* pathway) is the major pathway towards volatile terpenoids in plants including the light dependent release of isoprene from leaves (willow, poplar, etc.) (Zeidler et al 1997). The most important aspect of the inhibitor experiments is the massive incorporation of [D$_2$]-DOX into sesquiterpenes and the apparently flexible management of resources between the cytosol and the chloroplast.

References

Alborn HT, Turlings TCJ, Jones TH, Stenhagen G, Loughrin JH, Tumlinson JH 1997 An elicitor of plant volatiles from beet armyworm secretion. Science 276:945–949

Arigoni D, Sanger S, Latzel C, Eisenreich W, Bacher A, Zenk M 1997 Terpenoid biosynthesis from 1-deoxy-D-xylulose in higher plants by intramolecular skeletal rearrangement. Proc Natl Acad Sci USA 94:10600–10605

Baldwin IT, Zhang ZP, Diab N et al 1997 Quantification, correlations and manipulations of wound-induced changes in jasmonic acid and nicotine in *Nicotiana sylvestris*. Planta 201: 397–404

Bischoff H, Angerbauer R, Bender J et al 1997 Cerivastatin — pharmacology of a novel synthetic and highly active HMG-CoA reductase inhibitor. Atherosclerosis 135:119–130

Boland W, Feng Z, Donath J, Gäbler A 1992 Are acyclic C_{11} and C_{13} homoterpenes plant volatiles indicating herbivory? Naturwissenschaften 79:368–371

Boland W, Hopke J, Donath J, Nüske J, Bublitz F 1995 Jasmonic acid and coronatin induce odor production in plants. Angew Chem Int Ed Engl 34:1600–1602

Boland W, Gäbler A, Gilbert M, Feng Z 1998a Biosynthesis of C_{11} and C_{16} homoterpenes in higher plants: stereochemistry of the C–C-bond cleavage reaction. Tetrahedron 54:725–736

Boland W, Hopke J, Piel J 1998b Induction of plant volatile biosynthesis by jasmonates. In: Schreier P, Herderich M, Humpf HU, Schwab W (eds) Natural product analysis, chromatography–spectroscopy–biological testing. Viehweg Verlag, Braunschweig/ Wiesbaden, p 255–269

Cane DE, Ha HJ, MacIlwaine DB, Oascoe KO 1990 The synthesis of (3R)-nerolidol. Tetrahedron Lett 31:7553–7556

Chappell J 1995 Biochemistry and molecular biology of isoprenoid biosynthetic pathway in plants. Annu Rev Plant Physiol Plant Mol Biol 46:521–547

Dicke M, Sabelis M 1992 Costs and benefits of chemical information conveyance: proximate and ultimate factors. In: Roitberg BD, Isman MB (eds) Insect chemical ecology: an evolutionary approach. Chapman & Hall, New York, p 122–155

Dicke M, Sabelis MW, Takabayashi J, Bruin J, Posthumus MA 1990 Plant strategies of manipulating predator–prey interactions through allelochemicals: prospects for application in pest control. J Chem Ecol 16:3091–3118

Doares SH, Naváez-Vásques J, Conconi A, Ryan CA 1995 Salicylic acid inhibits synthesis of proteinase inhibitors in tomato leaves induced by systemin and jasmonic acid. Plant Physiol 108:1741–1746

Donath J, Boland W 1995 Biosynthesis of acyclic homoterpenes: enzyme selectivity and absolute configuration of the nerolidol precursor. Phytochemistry 39:785–790

Falkenstein E, Groth B, Mithöfer A, Weiler EW 1991 Methyl jasmonate and α-linolenic acid are potent inducers of tendril coiling. Planta 185:316–322

Farmer EE, Ryan CA 1990 Interplant communication: airborne methyl jasmonate induces synthesis of proteinase inhibitors in plant leaves. Proc Natl Acad Sci USA 87:7713–7716

Feussner I, Kindl H 1994 Particulate and soluble lipoxygenase isoenzymes — comparison of molecular and enzymatic properties. Planta 194:22–28

Feussner I, Wasternack C, Kindl H, Kühn H 1995 Lipoxygenase catalysed oxygenation of storage lipids is implicated in lipid mobilization during germination. Proc Natl Acad Sci USA 92:11849–11853

Gäbler A, Boland W, Preiss U, Simon H 1991 Stereochemical studies on homoterpene biosynthesis in higher plants: phylogenetic, ecological and mechanistic aspects. Helv Chim Acta 74:1773–1789

Gershenzon J, Croteau R 1993 Terpenoid biosynthesis: the basic pathway and formation of monoterpenes, sesquiterpenes, and diterpenes. In: Moore T Jr (ed) Lipid metabolism in plants. CRC Press, Boca Raton, FL, p 339–388

Giner JL, Jaun B, Arigoni D 1998 Biosynthesis of isoprenoids in *Escherichia coli*: the fate of the 3-H and 4-H atoms of 1-deoxy-D-xylulose. Chem Commun 1857–1858

Hamberg M 1988 Biosynthesis of 12-oxo-10,12(Z)-phytodienoic acid: identification of an allene oxide cyclase. Biochem Biophys Res Commun 156:543–550

Harrewijn P, Minks AK, Mollema C 1995 Evolution of plant volatile production in insect–plant relationships. Chemoecology 5/6:55–73

Hopke J 1997 Die Rolle der Jasmonsäure und ihrer Metaboliten im Signalstoffwechsel höherer Pflanzen. PhD thesis, University of Bonn, Germany

Hopke J, Donath J, Blechert S, Boland W 1994 Herbivore-induced volatiles: the emission of acyclic homoterpenes from leaves of *Phaseolus lunatus* and *Zea mays* can be triggered by β-glucosidase and jasmonic acid. FEBS Lett 352:146–150

Ichihara A, Shiraishi K, Sato H et al 1977 The structure of coronatine. J Am Chem Soc 99: 636–637

Kite GC 1995 The floral odor of *Arum maculatum*. Biochem System Ecol 23:343–354

Knudsen JT, Tollsten L, Bergström LG 1993 Floral scents — a checklist of volatile compounds isolated by headspace techniques. Phytochemistry 33:253–280

Kramell R, Atzorn R, Schneider G et al 1995 Occurrence and identification of jasmonic acid and its amino acid conjugates induced by osmotic stress in barley tissue. J Plant Growth Regul 14:29–36

Krumm T 1998 Aminosäurekonjugate in pflanzlichen Signalschienen Synthese und biologische Aktivität. PhD thesis, University of Bonn, Germany

Krumm T, Bandemer K, Boland W 1995 Induction of volatile biosynthesis in the lima bean (*Phaseolus lunatus*) by leucine- and isoleucine conjugates of 1-oxo and 1-hydroxyindan-4-carboxylic acid: evidence for amino acid conjugates of jasmonic acid as intermediates in the octadecanoid signaling pathway. FEBS Lett 377:523–529

Lange BM, Wildung MR, McCaskill D, Croteau R 1998 A family of transketolases that directs isoprenoid biosynthesis via a mevalonate-independent pathway. Proc Natl Acad Sci USA 95:2100–2104

Langenheim JH 1994 Higher plant terpenoids — a phytocentric overview of their ecological roles. J Chem Ecol 20:1223–1280

Lichtenthaler HK 1998 The plants' 1-deoxy-D-xylulose-5-phosphate pathway for bio-synthesis of isoprenoids. Fett Lipid 100:128–138

Lichtenthaler HK, Rohmer M, Schwender J 1997 Two independent biochemical pathways for isopentenyl diphosphate and isoprenoid biosynthesis in higher plants. Physiol Plant 101: 643–652

McCloud ES, Baldwin IT 11997 Herbivory and caterpillar regurgitants amplify the wound-induced increases in jasmonic acid but not nicotine in *Nicotiana sylvestris*. Planta 203:430–435

Midgely JJ, Bond WJ 1991 How important is biological pollination and dispersal to the success of the angiosperms? Phil Trans R Soc London B Biol Sci 333:209–215

Miksch M, Boland W 1996 Airborne methyl jasmonate stimulates the biosynthesis of furancoumarins in the leaves of celery plants (*Apium graveolens*). Experientia 52:739–743

Nüske J, Bublitz F 1993 *In vitro* coronatine production by several *Pseudomonas syringae* PV glycinea isolates. J Basic Microbiol 33:241

Peña-Cortés H, Albrecht T, Prat S, Weiler EW, Willmitzer L 1993 Aspirin prevents a wound-induced gene expression in tomato leaves by blocking jasmonic acid biosynthesis. Planta 191:123–128

Piel J, Boland W 1997 Highly efficient and versatile synthesis of isotopically labelled 1-deoxy-D-xylulose. Tetraherdron Lett 28:6397–6390

Piel J, Atzorn R, Gäbler R, Kühnemann F, Boland W 1997 Cellulysin from the plant parasitic fungus *Trichoderma viride* elicits volatile biosynthesis in higher plants via the octadecanoid signaling cascade. FEBS Lett 416:143–148

Piel J, Donath J, Bandemer K, Boland W 1998 Mevalonate-independent biosynthesis of terpenoid volatiles in plants: induced and constitutive emitted volatiles. Angew Chem Int Ed Engl 37:2478–2481

Rohmer M, Knani M, Simonin P, Sutter P, Sahm H 1993 Isoprenoid synthesis in bacteria — a novel pathway for the early steps leading to isopentenyl diphosphate. Biochem J 295:527–524

Sembdner G, Parthier B 1993 The biochemistry and the physiological and molecular actions of jasmonates. Annu Rev Plant Physiol Plant Mol Biol 44:569–589

Shulaev V, Silverman P, Raskin I 1997 Airborne signaling by methyl salicylate in plant pathogen resistance. Nature 385:718–721

Stanjek V, Herhaus C, Ritgen U, Boland W, Städler E 1998 Changes in the leaf surface chemistry of *Apium graveolens* (Apiaceae) stimulated by jasmonic acid and perceived by a specialist insect. Helv Chim Acta 80:1408–1420

Tamogami S, Rakwal R, Kodama O 1997 Phytoalexin production by amino acid conjugates of jasmonic acid through induction of naringenin-7-O-methyltransferase, a key enzyme in phytoalexin biosynthesis in rice (*Oryza sativa* L.). FEBS Lett 401:239–242

Turlings TCJ, Tumlinson JH, Lewis WJ 1990 Exploitation of herbivore-induced plant odours by host-seeking wasps. Science 250:1251–1253

Vick B, Zimmermann D 1984 Biosynthesis of jasmonic acid by several plant species. Plant Physiol 75:458–461

Wasternack C, Ortel B, Miersch O et al 1998 Diversity in octadecanoid-induced gene expression of tomato. J Plant Physiol 152:345–352

Weiler EW 1997 Octadecanoid-mediated signal transduction in higher plants. Naturwissenschaften 84:340–349

Yoshihara T, Greulich F 1999 Biosynthesis of jasmonoids and their function. In: Barton DRH, Nakanishi K (eds) Comprehensive natural product chemistry. Pergamon Press, Oxford, p 117–138

Zeidler JG, Lichtenthaler MK, May HU, Lichtenthaler FW 1997 Is isoprene emitted by plants synthesized via the novel isopentenyl pyrophosphate pathway? Z Naturforsch Sect C J Biosci 52:15–23

Zhang Z, Yang MJ, Pawliszyn J 1994 Solid-phase microextraction. Z Analyt Chem 66:844–853

DISCUSSION

Tumlinson: How did you apply these compounds to the lima beans?

Boland: In the same way you did! We dissolve the compounds in water and apply them to the petiole. The alternative methodology of spraying aqueous suspensions onto the leaf surface will work if the compounds are very lipophilic.

Takabayashi: Is the volicitin β-oxidized to hydroxyjasmonic acid?

Boland: This is a question we haven't been able to answer experimentally. But if volicitin were to be β-oxidized and finally produce hydroxyjasmonic acid, we already know that this compound is inactive. Hydroxylated JA derivatives don't induce volatile biosynthesis.

Beale: All your compounds are amides. These amides seem to be active regardless of the structure of the molecule. Is there any evidence to suggest that these amides are acting because they are blocking the degradation of endogenous JA? If this were the case, JA could not be turned over to its own amide and removed from the system. Is there any increase in endogenous JA?

Boland: No, there is no elevation of endogenous JA. What we do not know is whether specific receptors for these amides exist, or whether these amides just reside in the receptors which simply happen to be there. Could the indanones bind to the jasmonate receptors, and could it mean that the coronatin binds to the PDA receptor?

Scheel: That was exactly my question: are there separate receptors? One way to test this would be through competition experiments. Have you done this?

Boland: There is one competition experiment in the literature, involving a non-hydrolyzable conjugate of JA. Claus Wasternack used this in his barley system and found that if he pre-incubates the barley system with these non-hydrolyzable conjugates, it no longer responds to a subsequent jasmonate treatment (Wasternack et al 1998). These compounds apparently block something. Interestingly, he demonstrated which kind of compounds are up-regulated and which are down-regulated, and after pre-incubation he often saw exactly the opposite effects.

Scheel: You have a number of amino acid conjugates that are not active: do some of them act antagonistically?

Boland: This has not been checked.

Schultz: If you were to think like a pathogen or an insect, could you design the appropriate competitors for some of these signalling molecules to block a plant response?

Boland: In theory, yes. For example, we know that aphids introduce auxins into the plant. A possible strategy would be to introduce something which mimics endogenous signalling effects. Volicitin might be such a component, and coronatin surely is such a component. In that view we have a number of organisms which apparently produce compounds like this which change the physiology of the plant.

Schultz: Along those lines, don't different plant pathogens offer different substrates for the octadecanoid pathway? For example, arachidonic acid from the fungus, late blight, is a substrate for lipoxygenase in potato.

Boland: It could be so, but this has not been checked in detail. We are currently doing experiments with arachidonic acid to see whether or not volatiles are induced, but I don't know the results.

Schultz: I'm curious about the availability to these organisms of opportunities to manipulate these pathways.

Beale: It is rather strange that arachidonic acid is such a good elicitor. Does it enter the lipoxygenase pathway?

Boland: We can't say that for sure. One could imagine that it is converted to hydroperoxides, some of which could at least in principle be converted by the enzymes of the octadecanoid pathway. It is not known whether these compounds are further metabolized through cyclopentanyl derivatives, but there is, indeed, speculation in the literature (Stelzig et al 1983, Miller et al 1994).

Takabayashi: When JA derivatives are applied, some plants produce methyl salicylate. This may mean that salicylic acid is produced in the plant tissue. Is it possible that JA induces the production of salicylic acid that elicits the same defence signalling pathway?

Boland: This has not been checked. However, we do know that salicylate inhibits the action of JA.

Firn: You mentioned plant–plant signalling. I think we should be very cautious about this: there are severe theoretical barriers which make it unlikely to be operating in the real world (Firn & Jones 1995).

Boland: At least in principle it has been shown that plant–plant signalling could exist. Positive effects have been shown for ethylene, methyl salicylate and methyl jasmonate.

Firn: Volatile chemicals can induce effects in the laboratory, but that's all one can say.

Boland: On the other hand, if you can show that it is possible in principal, I would be surprised if it didn't exist in nature.

Pickett: What exactly is your philosophical argument against plant–plant signalling, Richard Firn?

Firn: The worst problem is that if the plant uses a generic chemical which many plants are making (such as ethylene, methyl salicylate or methyl jasmonate), it loses any signalling specificity. If you are a plant adjacent to a plant being attacked by a host-specific insect, and the attacked plant gives off this generic chemical and you respond to that, you are inducing defences which can be of no value, because that insect is not going to attack you. There is a high cost with no benefit. There is also a second serious theoretical obstacle. If you are producing low levels of a generic signalling substance and you are using that same compound to get information from another plant, unless the signalling plant can elevate the concentration of the chemical in the recipient above the level the recipient receives from its own background production, there can be no useful signal. In an open system, where massive dilution of the signal will occur between the signalling plant and the recipient, this is very hard to achieve with any reliability. The sums do not add up (see *http://www-users.york.ac.uk/~drf1/tt.htm*). These two theoretical objections seem to me to be severe.

Pickett: One can imagine ecological situations where there could be a benefit from this signalling, though. As Wilhelm Boland says, if you can demonstrate this in the laboratory, there is the chance that it is working naturally.

Firn: But the demonstrations in the laboratory have only been achieved to date by using high concentrations to get the signal-to-noise ratio into a workable range.

Pickett: It can also work both ways. I don't want to break any confidences, but I know from Jan Pettersson (Uppsala Agricultural University) that adjacent plants can actually disadvantage the plant that is purportedly receiving the signal (personal communication). So this is an allelopathic effect, which could be advantageous.

Firn: I'm open to experimental evidence which will suggest that such interactions could work in the field, but I would also welcome some more convincing evolutionary models as to the benefits of such signalling systems.

Dicke: The same argument could be held against using aggregation pheromones in insects: there, the receiver is also the producer of the same compounds. In fact, what you're saying is that if you use communication systems where the receiver is exposed to a compound that it also produces, it can't work.

Firn: I don't think that follows. I'm not an expert in aggregation pheromones, but presumably they are species specific. The crux of the problem I have is the total noise in the system. If you get the signal-to-noise ratio above that in a particular circumstance, you can then have a workable system. I suspect that insects improve the signal-to-noise ratio by processing temporal information—an option not available to plants.

Boland: I don't believe that there is real noise. Consider, for example, ethylene, which is the most potent candidate in this scenario. It is emitted as a burst. So there is a timing of the signal, a defined concentration, and so on: that is not really noise. There is a very strong ethylene burst after pathogen infection or herbivory.

Schultz: I think the major barrier here is technical. Over the last day and a half most of the conversations I have had at this meeting have focused on sampling. In particular, we have been asking whether we will ever be able to sample at a level that will allow us to answer any of these questions outdoors. I would certainly prefer to leave the evolutionary arguments here to the point where we know whether or not this was a reasonable phenomenon outdoors. One of my colleagues is a chemical engineer who is interested in trying to predict the concentration of things like methyl jasmonate in open air. He estimated that in a realistic setting, trying to detect methyl jasmonate moving between two tomato plants is like trying to find two people in the city of Tokyo with no phonebook or any other aid. Technically, it's a difficult problem. But we had the same problem with insect sex pheromones until we had GC and EAG. I think it's just a matter of time.

Lamb: What would constitute proof? Presumably you would need to show that a mutant that can't respond to jasmonate is under an ecological disadvantage relative to wild-type plants because it can't process incoming signals from surrounding plants.

Pickett: That would be suitable. Also, if there are plants standing in the field which are different merely from their association with other plants, then there must be some means of resourcing these mutually active compounds. However it is necessary to demonstrate that there is an effect by the compounds at their natural levels. As you know, we at Rothamsted are trying to use insects to plug into this system, because it is otherwise difficult to identify those chemicals from the plant that could be affecting the adjacent plants.

Firn: But the big difference is that the insects have a much better signal processing system because they have a brain capable of comparing the outputs of many individual sensors and they can gain information from temporal changes.

Pickett: However, as soon as it is demonstrated that a signal can pass into the plant from the air, even under artificial circumstances as with the use of a wind tunnel, then you raise the intriguing possibility of externally controlling plant gene expression. There is evidence working in both directions within very closely related plants from Jan Pettersson, which is in press. I agree with Jack Schultz: if we improve our techniques for actually seeing what is going on, then we may well find out there's a lot more there.

Scheel: Not only are the endogenous levels of a certain compound important, but also its compartmentation. When you're talking about internal jasmonate, you don't refer to any compartmentalization. Do you know where it is localized?

Boland: When we use labelled jasmonate, we see that the portion released to the atmosphere was only jasmonate we added exogenously: only the labelled component came out. The same result can be obtained if you use salicylate. It therefore appears that an excessive level of free material is immediately methylated and expelled to the atmosphere.

Scheel: I have a question on the point you raised that you need receptors for high molecular weight compounds. Are you sure that the receptor you were talking about really recognizes an enzyme in the cell lysate, or could it also be that such an enzyme releases low molecular weight compounds from the plant cell wall which are then recognized?

Boland: We tried to get pure proteins; we are down to about four proteins and they're still active. These are definitely proteins — the components are heat labile, for example. Interestingly, cellulysin acts in many plants. We have checked tobacco, lima bean, black alder and corn. In each, this enzyme cocktail induces volatiles, so it must be a very general principle, and it would be good to know what kind of secondary metabolites follow the action of the cell damage or lysis.

Schultz: I have a point of curiosity. At least one insect of which I know includes methyl jasmonate as a component in its sex pheromone. Is anything known about whether this is an acquisition from the plant, or is there a biosynthetic way to get that?

Boland: There are no biosynthetic experiments with that insect (*A mauris ochlea*). It has been published by Petty et al (1977). No further work has been done on that as far as I am aware.

References

Firn RD, Jones CG 1995 Plants may talk but can they hear? Trends Ecol Evol 10:371

Miller JS, Nguyen T, Stanley-Samuelson DW 1994 Eicosanoids mediate insect nodulation responses to bacterial infections. Proc Natl Acad Sci USA 91:12418–12422

Petty RL, Boppre M, Schneider D, Melnwald J 1977 Identifcation and localization of volatile hairpencil components in male *Amauris ochlea* butterflies (Danaidae). Experientia 33: 1324–1326

Stelzig DA, Allen RD, Bhatia SK 1983 Inhibition of phytoalexin synthesis in arachidonic acid-stressed potato tissue by inhibitors of lipoxygenase and cyanide resistant respiration. Plant Physiol 72:746–749

Wasternack C, Ortel B, Miersch O et al 1998 Diversity in octadecanoid-induced gene expression of tomato. J Plant Physiol 152:345–352

Diversity and variability of terpenoid defences in conifers: molecular genetics, biochemistry and evolution of the terpene synthase gene family in grand fir (*Abies grandis*)

Jörg Bohlmann* and Rodney Croteau

Institute of Biological Chemistry, Washington State University, Pullman, WA 99164-6340, USA

Abstract. This review focuses on the molecular genetics, biochemistry and evolution of terpenoid synthases relevant to terpenoid defences in conifers. In grand fir (*Abies grandis*) biosynthesis of terpenoids of the three classes of monoterpenes, sesquiterpenes and diterpenes is inducible by stem wounding at the level of gene activation and increase of enzyme activity of the respective terpene synthases. The monoterpene, sesquiterpene and diterpene synthases utilize prenyl diphosphates of appropriate size as substrates to generate the large diversity of carbon skeletons characteristic of the terpenoid resin of conifers. A large and diverse gene family of grand fir terpene synthases has been cloned and cDNAs are actively expressed in *Escherichia coli* for enzyme characterization. The monophyletic group of grand fir monoterpene, sesquiterpene and diterpene synthases represents both constitutively expressed and inducible genes encoding single product and multiple product enzymes. Several events of gene duplication and functional specialization of new synthases occurred during the evolution of terpenoid biosynthesis in grand fir, and gave rise to the enormous diversity and variability of this ancient and successful plant defence against herbivores and pathogens. The review concludes with a perspective of the biotechnological applications of terpenoid synthases for the genetic engineering of agricultural crops and forest trees.

1999 Insect–plant interactions and induced plant defence. Wiley, Chichester (Novartis Foundation Symposium 223) p 132–149

Terpenoid pathway and the role of terpenoid synthases

Terpenoids are by far the largest group of plant products, possessing a wide variety of physiological and ecological roles. The pathways of monoterpene,

---

*Present address: Max-Planck-Institut für Chemische Ökologie, Tatzendpromenade 1a, D-07745 Jena, Germany.

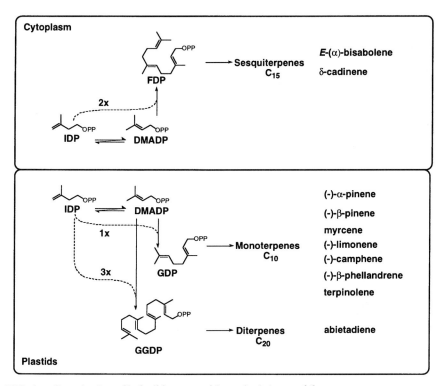

FIG. 1.   Organization of inducible terpene biosynthesis in grand fir.

sesquiterpene and diterpene biosyntheses are conveniently divided into several stages (Fig. 1). The first encompasses the synthesis of isopentenyl diphosphate (IDP), isomerization to dimethylallyl diphosphate (DMADP), prenyltransferase-catalyzed condensation of these two C_5 units to geranyl diphosphate (GDP), and the subsequent 1'–4 additions of IDP to generate farnesyl diphosphate (FDP) and geranylgeranyl diphosphate (GGDP) (Ogura & Koyama 1999). In the second stage, the prenyl diphosphates undergo a range of cyclizations based upon variations on the same mechanistic theme to produce the parent skeletons of each class. Thus, GDP (C_{10}) gives rise to monoterpenes (Wise & Croteau 1999), FDP (C_{15}) to sesquiterpenes (Cane 1999), and GGDP (C_{20}) to diterpenes (MacMillan & Beale 1999). These transformations catalyzed by the terpenoid synthases (cyclases) may be followed by a variety of redox modifications of the parent skeletal types to produce the many thousands of different terpenoid metabolites of the essential oils, turpentines and resins of plant origin (Connolly & Hill 1991). Terpene synthases generate the enormous diversity of carbon skeletons characteristic of terpenoids.

Most of these natural products are cyclic, and many contain multiple ring systems, the basic structures of which are determined by the highly specific terpenoid synthases. The terpenoid synthases may be involved in the regulation of pathway flux since they operate at metabolic branch-points and catalyse the first committed steps leading to the various terpene classes (Gershenzon & Croteau 1993). The remarkably similar characteristics of the monoterpene, sesquiterpene and diterpene synthases (Alonso & Croteau 1993) are certainly related to the fact that these enzymes carry out similar electrophilic cyclizations (Wise & Croteau 1999, Cane 1999, MacMillan & Beale 1999) involving common steps, i.e. the generation, transformation and stabilization of highly reactive carbocations and their ultimate quenching by deprotonation or nucleophile capture. The unique features of each individual cyclization relate to the precise means by which each synthase enforces conformation on the substrate and intermediates to effect a particular reaction channel while protecting such reactive species from premature termination. Although the size and shape of the active sites must differ (Starks et al 1997), the basic means by which these enzymes enforce regiochemistry and stereochemistry of product formation are also probably very similar. This similarity in function is reflected in similarity in primary structure of all of the terpene synthases of plant origin (Bohlmann et al 1998a).

Molecular cloning of plant terpenoid synthases

In the last few years, the molecular cloning of terpene synthases has contributed much to our understanding of terpenoid biosynthesis in plants. Cloning strategies and sequence analysis of terpenoid synthases have recently been reviewed (Bohlmann et al 1998a). Almost 40 plant terpenoid synthases have now been cloned and many of the cDNAs have been expressed in *Escherichia coli* for characterization of the recombinant enzymes. Sequence comparison and phylogenetic reconstruction places all typical terpenoid synthases of secondary metabolism into the subfamilies *Tpsa*, *Tpsb* and *Tpsd* of the large *Tps* gene family (Bohlmann et al 1998a) (Fig. 2). For each of these subfamilies the first cloned genes were obtained using reverse genetic strategies based on the purified proteins. Subsequently, similarity-based PCR strategies were employed for the cloning of related enzymes. Similarity-based cloning of members of the *Tpsd* subfamily of gymnopsperm terpene synthases was possible after purification and protein-based cloning of the grand fir diterpene synthase abietadiene synthase (Stofer Vogel et al 1996). Sequence comparison of abietadiene synthase with a few members of other subfamiles allowed the similarity based cloning using PCR of seven monoterpene synthases and three sesquiterpene synthases from grand fir (Bohlmann et al 1997, 1998b, 1999, Steele et al 1998a), as well as the PCR-based cloning of taxadiene synthase from *Taxus brevifolia* (Wildung & Croteau 1996).

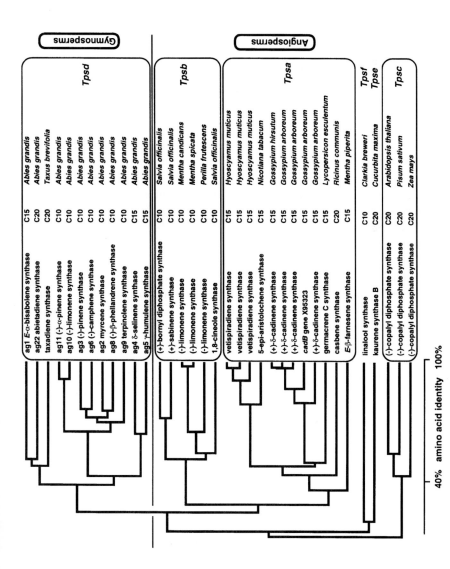

FIG. 2. Sequence relatedness of the plant *Tps* gene family (for references see Bohlmann et al 1998a).

The *Tpsd* subfamily of gymnosperm terpene synthases now represents the largest and, with respect to catalytic functions of its members, the most diverse monophyletic group of functionally characterized terpene synthases (Bohlmann et al 1998a).

In addition to reverse genetic strategies and similarity-based PCR cloning of terpene synthases, efforts to sequence the complete genomes of model plants such as *Arabidopsis*, rice or corn will inadvertently reveal new terpene synthase-like genes (Aubourg et al 1997, Bevan et al 1998). Genome sequencing can lead to the identification of new terpene synthases, when the corresponding candidate cDNAs are actively expressed in a suitable host for functional characterization of the enzymes. It is currently not possible to predict the specific catalytic activity of a new terpene synthase-like gene based on sequence analysis only. Studies with grand fir monoterpene cyclases demonstrated that synthases as closely related as 91% identity at the amino acid level can encode biochemically distinct terpene synthases, such as (−)-limonene synthase (Bohlmann et al 1997) and (−)-α-pinene synthase (Bohlmann et al 1999). However, synthases which share less than 30% amino acid identity, can catalyse the same cyclization reaction, such as (−)-limonene synthase from mint (Colby et al 1993) and (−)-limonene synthase from grand fir (Bohlmann et al 1997).

Although terpene synthases of all three classes of monoterpene, sesquiterpene and diterpene synthases have now been cloned from gymnosperms and angiosperms, no sequence of a terpene synthase of secondary metabolism has yet been described for a member of the monocotyledonous class of angiosperms. Efforts to clone terpene synthases from corn by PCR using a large number of combinations of degenerate oligonucleotides, which have previously successfully been employed for similarity-based cloning of terpene synthases from other species, failed to yield specific gene fragments (J. Bohlmann, unpublished results). These negative results can imply that monoterpene and sesuiterpene synthases in corn are not members of those *Tps* subfamilies described until now, but may represent phylogenetically distant enzymes.

Heterologous expression of plant terpenoid synthases in *E. coli*

Heterologous expression as an active recombinant protein and identification of the enzymatic reaction products formed with the possible C_{10}, C_{15} or C_{20} prenyl diphosphate substrates are required for functional characterization of cloned terpenoid synthases. Terpenoid synthases are operationally soluble enzymes localized to the cytosol (sesquiterpene synthases) or plastids (monoterpene synthases and diterpene synthases) which can be functionally expressed in *E. coli* using standard plasmid expression vectors. However, in some cases, the N-terminal transit peptides of these synthases, which target the nuclear-encoded preproteins to the plastids for proteolytic processing to the mature forms, promote formation of inclusion bodies when expressed in *E. coli*. High yield expression of soluble monoterpene synthases can be achieved by truncation of the cDNAs to remove the targeting sequences (Williams et al 1998, Bohlmann et al 1999). Another commonly encountered problem with expression of plant terpenoid synthases in *E. coli* relates to the frequency of Arg residues which use rare tRNAs in the prokaryotic host. Coexpression of the terpenoid synthase cDNA with the required tRNA can eliminate translational difficulty and yield high level expression of active recombinant enzymes (Steele et al 1998a, Williams et al 1998, Bohlmann et al 1999).

Since most terpenoid synthases are present in plants at low levels and are not easily purified, heterologous expression of cloned synthases has played an important role in obtaining high yield of pure proteins for antibody preparation or crystallization for structural analysis. cDNA cloning and functional expression has further provided unambiguous proof that many terpenoid synthases form multiple products in fixed ratios. For example, recombinant limonene synthase converts GDP to $(-)$-limonene, α-pinene, β-pinene and myrcene in a ratio identical to the native enzyme (Rajaonarivony et al 1992a, Colby et al 1993), and grand fir $(-)$-pinene synthase, in both native and recombinant form, produces $(-)$-α-pinene and $(-)$-β-pinene at a ratio of 2:3 (Lewinsohn et al 1992, Bohlmann et al 1997). Expression of the cDNAs for grand fir monoterpene synthases has provided evidence that the complex turpentine mixture produced by this conifer (Lewinsohn et al 1993) is formed by a family of single product and multiproduct enzymes encoded by closely related genes (Bohlmann et al 1997, 1999). Single product and multiproduct sesquiterpene synthases also exist in grand fir. Thus, δ-selinene synthase and γ-humulene synthase form, respectively, 34 and 52 different sesquiterpenes (Steele et al 1998a), while a third synthase from grand fir produces only (E)-α-bisabolene (Bohlmann et al 1998b).

Testing with different prenyl diphosphate substrates of heterologously expressed enzymes indicates that substrate specificities have evolved differently in the three classes of terpenoid synthases. All monoterpene synthases and diterpene synthases reveal strict substrate specificity in accepting only GDP or GGDP, respectively, as might be expected for the purpose of controlling flux through these plastidial pathways for isoprenoid metabolism. Selectivity in the case of monoterpene synthases can be explained as a result of size exclusion for FDP or GGDP, but other refinements in active site structure are required to exclude GDP and FDP as substrates for diterpene synthases. Interestingly, several sesquiterpene synthases will accept GDP (but not GGDP) as an alternate, but inefficient, substrate for the formation of the simple olefin limonene (Crock et al 1997, Colby et al 1998, Steele et al 1998a, Bohlmann et al 1998b). Under physiological conditions, the cytosolic sesquiterpenoid synthases are unlikely to encounter GDP, which arises in plastids (Gershenzon & Croteau 1993), and thus there is no evolutionary pressure for discrimination against this adventitious substrate.

General and specific structural features of terpene synthases

Terpenoid synthase cDNAs encode proteins of 550–850 amino acids (Bohlmann et al 1998a), in agreement with observed native molecular weights of 50–100 kDa. In general, monoterpene synthases are between 600 and 650 amino acids in length, and are larger by 50–70 amino acids than sesquiterpene synthases. This difference is due largely to the N-terminal transit peptides required for plastidial targeting of monoterpene synthases. While lacking similarity in primary structure, the targeting sequences are characterized by a high content of serine and threonine and a low number of acidic residues (Williams et al 1998). N-terminal deletion has demonstrated that residues upstream of a highly conserved RR motif are not required for monoterpene synthase activity (Williams et al 1998, Bohlmann et al 1999). Since sequence similarity among these synthases is significant only downstream of the RR motif, this element may define the approximate N-terminus of mature processed proteins. The gymnosperm diterpene synthases, abietadiene synthase (Stofer Vogel et al 1996) and taxadiene synthase (Wildung & Croteau 1996) are approximately 210 amino acids longer than even monoterpene synthases. The difference in length is accounted for by an additional internal element that is conserved in sequence and position among all diterpene synthases regardless of cyclization mechanism (Bohlmann et al 1998a), with the exception of casbene synthase (Mau & West 1994). Interestingly, this sequence is also conserved in *Clarkia* linalool synthase (Dudareva et al 1996), a monoterpene synthase, and in the grand fir sesquiterpene synthase (E)-α-bisabolene synthase (Bohlmann et al 1998b). Until now, most terpenoid synthases have been cloned as cDNAs, and the structures of only a few terpene synthase genes have been described. Comparison of the genes for tobacco *epi*-aristolochene synthase, castor bean casbene synthase, *Hyoscyamus* vetispiradiene synthase, mint limonene synthase and a putative terpene synthase from *Arabidopsis* reveals a similar overall structure with six positionally conserved introns (Facchini & Chappell 1992, Mau & West 1994, Back & Chappell 1995, Yuba et al 1996, Aubourg et al 1997).

Recently, the first crystal structure of a plant terpene synthase was solved for tobacco *epi*-aristolochene synthase (Starks et al 1997), and revealed the protein to be composed entirely of α-helices with short connecting loops and turns, forming a two-layer α-barrel active site. At the entrance of the active site, the diphosphate of the prenylsubstrate moiety is complexed by two Mg^{2+} ions. One Mg^{2+} is coordinated by two conserved Asp residues and a Glu residue. Both aspartate residues are part of an absolutely conserved aspartate-rich motif, DDxxD

(Bohlmann et al 1998a). The second Mg^{2+} is coordinated by an Asp residue, a non-conserved Thr residue and by a well conserved Glu/Asp residue. In *epi*-aristolochene synthase, the side chain of an absolutely conserved Arg residue is thought to stabilize the negative charge of the diphosphate following ionization and to direct this species away from the active site to prevent reaction with carbocationic intermediates of the catalytic cycle. Aromatic residues, many of which are conserved (Bohlmann et al 1998a), may stabilize carbocationic intermediates. As modelled for *epi*-aristolochene synthase, the carboxyl of a conserved Asp residue could be involved in proton translocation in reaction intermediates. A tryptophan residue has been suggested as an unusual terminating proton acceptor in this case, in part because of the absence of a suitable His residue which has been implicated as the general base in other terpenoid synthases (Rajaonarivony et al 1992b, Lesburg et al 1997). Active site chemical modification studies implicate essential Cys, His and Arg residues in monoterpene synthases, with differential placement (or roles) in angiosperm and gymnosperm synthases (Rajaonarivony et al 1992b, Savage et al 1994, 1995). The structure of *epi*-aristolochene synthase resembles that of microbial synthases (Wendt et al 1997, Lesburg et al 1997), and modelling studies suggest that all terpenoid synthases bear the same overall fold structure. This common three-dimensional framework, when coupled to directed mutagenesis, provides the means of evaluating the functional relevance of conserved amino acids.

Evolution of the *Tps* gene family

On the basis of amino acid sequence relatedness of plant terpenoid synthases, the *Tps* gene family is subdivided into six subfamilies (Fig. 2) (Bohlmann et al 1997, 1998a), designated *Tpsa* through *Tpsf*. The terpenoid synthases of primary metabolism, (−)-CDP synthase (*Tpsc*) and kaurene synthase B (*Tpse*), are only distantly related to those of secondary metabolism, including members of subfamilies *Tpsa*, *Tpsb* and *Tpsd*. However, all plant terpene synthases share a common evolutionary origin (Bohlmann et al 1998a). Terpenoid synthases of secondary metabolism constitute the most extensively studied *Tps* subfamilies, including *Tpsa*, *Tpsb*, *Tpsd* and the distant and possibly ancient *Tpsf* branch containing linalool synthase. The angiosperm monoterpene synthase subfamily *Tpsb*, representing members of the Lamiaceae, is clearly distinct from angiosperm sesquiterpene synthases (*Tpsa*) and from gymnosperm monoterpene synthases (*Tpsd*). Sequence comparison and phylogenetic reconstruction consistently reveal that gymnosperm monoterpene, sesquiterpene and diterpene synthases represented by genes from grand fir and *Taxus brevifolia* are more closely related to each other than they are to their counterparts of angiosperm origin (Bohlmann et al 1998a). Thus, the gymnosperm terpenoid synthases form a separate branch designated as the *Tpsd* group. The pattern of bifurcation of gymnosperm and angiosperm terpenoid synthases from a common ancestor implies independent functional specialization after separation of the angiosperm and gymnosperm lineages. Terpenoid synthases of the gymnosperm *Tpsd* subfamily show an enormous functional diversity. It can be generalized that many terpenoid synthases of secondary metabolism are encoded by multiple gene families, which arose by duplication and then provided the basis for diversification. In grand fir, at least seven very closely related genes (minimum of 65% identity) encode monoterpene synthases with distinct products (Bohlmann et al 1997, 1999). Because genes of secondary metabolism are not essential for growth and development, they may tolerate functional mutations leading to product diversity, and increased diversity in terpenoid chemistry may prove beneficial in ecological interactions with pathogens and herbivores.

Terpenoid defences in conifers and the role of terpenoid synthases in grand fir

Grand fir (*Abies grandis*) has been developed as a model system to study the complexity of terpenoid defence strategies and their molecular genetics and biochemistry in conifers. Chemical defences in conifers against herbivores and fungal pathogens involve at least three classes of terpenoids, monoterpenes, sesquiterpenes and diterpenes (Fig. 1). Monoterpenes are toxic to invading insects and furnish the volatile solvent for diterpene resin acids, which upon monoterpene evaporation harden to a mechanical barrier sealing the wound site (Johnson & Croteau 1987, Gijzen et al 1993). Inducible resin sesquiterpenoids could have similar functions as angiosperm phytoalexins in pathogen defence, while bisabolene-derived juvenile hormone analogues interfere with insect reproduction and development (Bowers 1991). Grand fir responds to mechanical wounding of stems, which mimics bark beetle attack, by the induced formation of defence-related monoterpenes, sesquiterpenes and diterpenes (Steele at al 1995). An enriched cDNA library from wound-induced sapling stems was employed for the cloning of 11 defence-related terpene synthases (Stofer Vogel et al 1996, Bohlmann et al 1997, 1998b, 1999, Steele et al 1998a), representing the largest group of synthases described now for any plant species (Bohlmann et al 1998a). All cloned grand fir terpene synthases are members of the *Tpsd* subfamily. Within this monophyletic group seven monoterpene synthases are most closely related among each other. Two sesquiterpene synthases, γ-humulene synthase and δ-selinene synthase also form a separate branch in the *Tpsd* group. A third grand fir sesquiterpene synthase, (E)-α-bisabolene synthase, is most closely related to gymnosperm diterpene synthases, abietadiene synthase and taxadiene synthase.

The cloned monoterpene synthases from grand fir represent single product enzymes, such as myrcene synthase, and enzymes which produce one major product and several minor products (Bohlmann et al 1997, 1999). Examples of the latter type are ($-$)-limonene synthase, β-phellandrene synthase, terpinolene synthase, ($-$)-α-pinene synthase and ($-$)-camphene synthase. A second ($-$)-pinene synthase forms both ($-$)-α-pinene and ($-$)-β-pinene. Monoterpene synthases are inducible at the level of mRNA accumulation by mechanical wounding of sapling stems (Bohlmann et al 1997, Steele et al 1998b).

In order to explore the level of biochemical variability of monoterpene biosynthesis in grand fir, 46 saplings were individually tested for constitutive and wound-induced terpene synthase enzyme activity, and the monoterpene products of these activities were identified (Katoh & Croteau 1998). Differences among non-clonal trees with respect to up-regulation and down-regulation of individual members of the monoterpene synthase group allowed definition of seven monoterpene-biosynthetic chemotypes, which reflect the possible biosynthetic diversity of monoterpene formation in a natural population. Independent expression of a family of monoterpene synthase genes, which encode enzymes with a high degree of product specificity, allows for the chemical diversity and variability of monoterpene defences. As a consequence, adaptation of herbivores to overcome the multigene and multi-component chemical defence is likely to be more difficult compared to a situation in which tree defence builds upon a single component or on a constant and invariable mixture of monoterpenoids. This concept can now be tested using genetic engineering of terpenoid defence patterns, which may provide new strategies for improving terpenoid defences in silviculture.

In contrast to the high degree of product specificity of grand fir monoterpene synthases, two cloned sesquiterpene synthases, δ-selinene synthase and γ-humulene synthase, represent multiproduct enzymes which form from FDP, respectively, 34 and 52 different products of enormous structural diversity (Steele et al 1998a). The two multiproduct sesquiterpene synthases are not inducible by wounding in grand fir (Steele et al 1998a). Differences in the level of expression of these enzymes have been observed with individual trees (J. Crock & R. Croteau, unpublished results). These most unusual multiproduct enzymes contribute directly to the chemical diversity of terpenoid composition based on the constitutive expression of only one or two genes. In evolutionary terms, these genes can provide the raw material for specialization of new defence genes.

Grand fir also expresses two wound-inducible sesquiterpene synthases, δ-cadinene synthase and (E)-α-bisabolene synthase (Steele et al 1998a). (E)-α-bisabolene synthase was cloned and, in contrast to the constitutive α-selinene synthase and γ-humulene synthase, this wound-inducible gene encodes for a highly specialized enzyme, which converts FDP exclusively into a single sesquiterpene product (Bohlmann et al 1998b). Feeding of labelled (E)-α-bisabolene to methyl jasmonate-induced grand fir cells results in incorporation into todomatuic acid, the precursor of the insect juvenile hormone analogue juvabione (Bohlmann et al 1998b). Thus, cloned (E)-α-bisabolene synthase provides a useful molecular tool for manipulating phyto-juvenile hormone biosynthesis in transgenic plants. Induced gene expression of the specialized (E)-α-bisabolene synthase is suggested to represent a second line of terpenoid defence against insect herbivores and pathogens. Accumulation of mRNA for monoterpene synthases increases within a few hours after wounding and reaches a maximum within two to four days (Steele et al 1998b). Diterpene synthase gene expression is similarly induced but the time-course is slightly delayed compared to monoterpene synthase induction. In contrast, (E)-α-bisabolene synthase mRNA shows its maximum level only two weeks after wounding (Bohlmann et al 1998b). These differences in the timing of induced gene expression are consistent with the different roles of terpenoid defence compounds. Thus, toxic monoterpenes are required immediately after insect attack, diterpene resin acids are mobilized somewhat later, while sesquiterpenoid juvenile hormone analogues are produced still later, presumably to interfere with insect reproduction and development should infestation of the host succeed.

Terpenoid defence in conifers not only involves metabolic redirection of constitutive processes and the induction of new pathways, but can also be accompanied by anatomical changes. In grand fir and in several spruce species, insect herbivory, fungal elicitation and mechanical wounding cause secondary resinosis in newly differentiated resin systems. Secondary or traumatic resin ducts develop in *Picea abies* upon bark beetle attack and inoculation with the blue stain fungus *Ceratocystis polonica* (E. Christiansen & V. Franceschi, personal communication). *Picea glauca* cambium cells, which normally form xylem, are programmed to yield axial traumatic resin ducts after attack by the white pine weevil *Pissodes strobi* (Alfaro 1995). Cloned terpene synthases provide probes to study the molecular and cellular events of insect and pathogen induced traumatic resin duct differentiation.

Perspective

Terpenoid secondary metabolites are recognized as signal molecules in many interactions of plants with other organisms, including competitors, beneficial insects, herbivores and microbial pathogens (Gershenzon & Croteau 1991). Terpenoid phytoalexins of the sesquiterpene and diterpene series serve as induced defences against fungal pathogens (Kuc 1995). Volatile terpenes serve as attractants for pollinators (Dobson 1993) and mediate certain forms of tritrophic interactions (Turlings et al 1995). Plant terpenoids can influence insect communication as pheromones or pheromone precursors (Pickett 1991), and can interfere with insect development (Bowers 1991). Terpenoid-mediated chemical communication between plants and insects reflects eons of co-evolution leading, in part, to both the diversity and specificity of terpene synthase catalysts. In the various forms of interaction of conifers with insects, terpenoids seem to be involved in many aspects of the herbivore's life cycle. Terpenoids serve as volatiles in host recognition, as pheromones in intraspecific and interspecific communication during host colonization, as feeding attractants or deterrents, as oviposition stimulants or deterrents, or as hormone analogues in insect development and reproduction. The many ecological concepts, which posit terpenoids as mediators of plant–pathogen and plant–insect interactions, can now be experimentally evaluated by means of transgenic approaches, which may in turn lead to new strategies for plant protection. Among the many biotechnological applications made possible by molecular cloning of terpenoid synthases, the genetic engineering of terpenoid-based defences in crop plants and trees is of particular interest, given the storehouse of possibilities and the high probability that these substances are active against pests not adapted to them. The terpenoid synthase multigene family provides the means for genetic engineering of plant populations with mosaics of chemical defences targeted to a range of insect ecological, physiological and developemental processes.

References

Alfaro RI 1995 An induced defense reaction in white spruce to attack by the white pine weevil, *Pissodes strobi*. Can J For Res 25:1725–1730

Alonso WR, Croteau R 1993 Prenyltransferases and cyclases. Methods Plant Biochem 9:239–260

Aubourg S, Takvorian A, Chéron A, Kreis M, Lecharny A 1997 Structure, organization and putative function of the genes identified within a 23.9 kb fragment from *Arabidopsis thaliana* chromosome IV. Gene 199:241–253

Back K, Chappell J 1995 Cloning and bacterial expression of a sesquiterpene cyclase from *Hyoscyamus muticus* and its molecular comparison to related terpene cyclases. J Biol Chem 270:7375–7381

Bevan M, Bancroft I, Bent E et al 1998 Analysis of 1.9 Mb of contiguous sequence from chromosome 4 of *Arabidopsis thaliana*. Nature 391:485–488

Bohlmann J, Steele CL, Croteau R 1997 Monoterpene synthases from grand fir (*Abies grandis*): cDNA isolation, characterization and functional expression of myrcene synthase, ($-$)-(4S)-limonene synthase, and ($-$)-(1S,5S)-pinene synthase. J Biol Chem 272:21784–21792

Bohlmann J, Meyer-Gauen G, Croteau R 1998a Plant terpenoid synthases: molecular biology and phylogenetic analysis. Proc Natl Acad Sci USA 95:4126–4133

Bohlmann J, Crock J, Jetter R, Croteau R 1998b Terpenoid-based defenses in conifers: cDNA cloning, characterization, and functional expression of wound-inducible (*E*)-α-bisabolene synthase from grand fir (*Abies grandis*). Proc Natl Acad Sci USA 95:6756–6761

Bohlmann J, Phillips M, Ramachandiran V, Katoh S, Croteau R 1999 cDNA cloning, characterization and functional expression of four new members of the *Tpsd* gene family from grand fir (*Abies grandis*). Arch Biochem Biophys, in press

Bowers WS 1991 Insect hormones and anti-hormones. In: Rosenthal GA, Janzen D (eds) Herbivores: their interaction with secondary plant metabolites, vol 1. Academic Press, San Diego, CA, p 431–456

Cane DE 1999 Sesquiterpene biosynthesis: cyclization mechanisms. In: Cane DE (ed) Comprehensive natural products chemistry, vol 2: Isoprenoids including carotenoids and steroids. Pergamon, Oxford, in press

Colby SM, Alonso WR, Katahira E, McGarvey DJ, Croteau R 1993 4S-limonene synthase from the oil glands of spearmint (*Mentha spicata*). cDNA isolation, characterization, and bacterial expression of the catalytically active monoterpene cyclase. J Biol Chem 268:23016–23024

Colby SM, Crock J, Dowdle-Rizzo B, Lemaux PG, Croteau R 1998 Germacrene C synthase from *Lycopersicon esculentum* cv VFNT cherry tomato: cDNA isolation, characterization, and bacterial expression of the multiple product sesquiterpene cyclase. Proc Natl Acad Sci USA 95:2216–2221

Connolly JD, Hill RA 1991 Dictionary of terpenoids. Chapman & Hall, London

Crock J, Wildung MR, Croteau R 1997 Isolation and bacterial expression of a sesquiterpene synthase cDNA clone from peppermint (*Mentha × piperita*, L.) that produces the aphid alarm pheromone (*E*)-β-farnesene. Proc Natl Acad Sci USA 94:12833–12838

Dobson HEM 1993 Floral volatiles in insect biology. In: Bernays E (ed) Insect–plant interactions, vol 5. CRC Press, Boca Raton, FL, p 47–81

Dudareva N, Cseke L, Blanc VM, Pichersky E 1996 Evolution of floral scent in *Clarkia*: novel patterns of S-linalool synthase gene expression in the *C. breweri* flower. Plant Cell 8:1137–1148

Facchini PJ, Chappell J 1992 Gene family for an elicitor-induced sesquiterpene cyclase in tobacco. Proc Natl Acad Sci USA 89:11088–11092

Gershenzon J, Croteau R 1991 Terpenoids. In: Rosenthal GA, Janzen D (eds) Herbivores: their interaction with secondary metabolites, vol 1. Academic Press, San Diego, CA, p 165–219

Gershenzon J, Croteau R 1993 Terpenoid biosynthesis: the basic pathway and formation of monoterpenes, sesquiterpenes, and diterpenes. In: Moore TS Jr (ed) Lipid metabolism in plants. CRC Press, Boca Raton, FL, p 339–388

Gijzen M, Lewinsohn E, Savage TJ, Croteau R 1993 Conifer monoterpenes. ACS Symp Ser 525:8–22

Johnson MA, Croteau R 1987 Biochemistry of conifer resistance to bark beetles and their fungal symbionts. ACS Symp Ser 325:67–91

Katoh S, Croteau R 1998 Individual variation in constitutive and induced monoterpene biosynthesis in grand fir. Phytochemistry 47:577–582

Kuc J 1995 Phytoalexins, stress metabolism, and disease resistance in plants. Annu Rev Phytopathol 33:275–297

Lesburg CA, Zhai G, Cane DE, Christianson DW 1997 Crystal structure of pentalenene synthase: mechanistic insights on terpenoid cyclization reactions in biology. Science 277:1820–1824

Lewinsohn E, Gijzen M, Croteau R 1992 Wound-inducible pinene cyclase from grand fir: purification, characterization, and renaturation after SDS-PAGE. Arch Biochem Biophys 293:167–173

Lewinsohn E, Savage TJ, Gijzen M, Croteau R 1993 Simultaneous analysis of monoterpenes and diterpenoids of conifer oleoresin. Phytochem Anal 4:220–225

MacMillan J, Beale M 1999 Diterpene biosynthesis. In: Cane DE (ed) Comprehensive natural products chemistry, vol 2: Isoprenoids including carotenoids and steroids. Pergamon, Oxford, in press

Mau CJD, West CA 1994 Cloning of casbene synthase cDNA: evidence for conserved structural features among terpenoid cyclases in plants. Proc Natl Acad Sci USA 91:8497–8501

Ogura K, Koyama T 1999 Isopentenyl diphosphate, isomerase and prenyltransferase. In: Cane DE (ed) Comprehensive natural products chemistry, vol 2: Isoprenoids including carotenoids and steroids. Pergamon, Oxford, in press

Pickett JA 1991 Lower terpenoids as natural insect control agents. In: Harborne JB, Tomas-Barberan FA (eds) Ecological chemistry and biochemistry of plant terpenoids. Clarendon Press, Oxford, p 297–313

Rajaonarivony JIM, Gershenzon J, Croteau R 1992a Characterization and mechanism of (4S)-limonene synthase, a monoterpene cyclase from the glandular trichomes of peppermint (Mentha×piperita). Arch Biochem Biophys 296:49–57

Rajaonarivony JIM, Gershenzon J, Miyazaki J, Croteau R 1992b Evidence for an essential histidine residue in 4S-limonene synthase and other terpene cyclases. Arch Biochem Biophys 299:77–82

Savage TJ, Hatch MW, Croteau R 1994 Monoterpene synthases of Pinus contorta and related conifers: a new class of terpenoid cyclase. J Biol Chem 269:4012–4020

Savage TJ, Ichii H, Hume SD, Little DB, Croteau R 1995 Monoterpene synthases from gymnosperms and angiosperms: stereospecificity and inactivation by cysteinnyl- and arginyl-directed modifying reagents. Arch Biochem Biophys 320:257–265

Starks CM, Back K, Chappell J, Noel JP 1997 Structural basis for cyclic terpene biosynthesis by tobacco 5-epi-aristolochene synthase. Science 277:1815–1820

Steele CL, Lewinsohn E, Croteau R 1995 Induced oleoresin biosynthesis in grand fir as a defense against bark beetles. Proc Natl Acad Sci USA 92:4164–4168

Steele CL, Crock J, Bohlmann J, Croteau R 1998a Sesquiterpene synthases from grand fir (Abies grandis): comparison of constitutive and wound-induced activities, and cDNA isolation, characterization, and bacterial expression of δ-selinene synthase and γ-humulene synthase. J Biol Chem 273:2078–2089

Steele C, Katoh S, Bohlmann J, Croteau R 1998b Regulation of oleoresinosis in grand fir (Abies grandis): differential transcriptional control of monoterpene, sesquiterpene, and diterpene synthase genes in response to wounding. Plant Physiol 116:1497–1504

Stofer Vogel B, Wildung MR, Vogel G, Croteau R 1996 Abietadiene synthase from grand fir (Abies grandis): cDNA isolation, characterization, and bacterial expression of a bifunctional diterpene cyclase involved in resin acid biosynthesis. J Biol Chem 271:23262–23268

Turlings TCJ, Loughrin JH, McCall PJ, Röse USR, Lewis WJ, Tumlinson JH 1995 How caterpillar-damaged plants protect themselves by attracting parasitic wasps. Proc Natl Acad Sci USA 92:4169–4174

Wendt KU, Poralla K, Schulz GE 1997 Structure and function of squalene cyclase. Science 277:1811–1815

Wildung MR, Croteau R 1996 A cDNA clone for taxadiene synthase, the diterpene cyclase that catalyzes the committed step of taxol biosynthesis. J Biol Chem 271:9201–9204

Williams DC, McGarvey DJ, Katahira EJ, Croteau R 1998 Truncation of limonene synthase preprotein provides a fully active 'pseudomature' form of this monoterpene cyclase and reveals the function of the amino-terminal arginine pair. Biochemistry 37: 12213–12220

Wise ML, Croteau R 1999 Monoterpene biosynthesis. In: Cane DE (ed) Comprehensive natural products chemistry, vol 2: Isoprenoids including carotenoids and steroids. Pergamon, Oxford, in press

Yuba A, Yazaki K, Tabata M, Honda G, Croteau R 1996 cDNA cloning, characterization, and functional expression of 4S-(−)-limonene synthase from Perilla frutescens. Arch Biochem Biophys 332:280–287

DISCUSSION

Boland: What is known about the folding and control of these multiproduct enzymes you described? One could imagine a scenario in which the enzyme just controls the formation of the cation and everything else is chemistry, depending on temperature. The other extreme scenario would be that you have a lot of subunits in the enzyme, each specific for a product? What is the reality?

Bohlmann: There is not much known about the structural details controlling the reaction mechanism of the multiproduct terpene synthases. The product profile does not depend on temperature or other non-enzymatic factors, but is determined by the structure of the active site of the enzymes. These enzymes are not composed of multiple subunits. The only plant sesquiterpene synthase that has been crystallized and the active site structure analysed, is the tobacco *epi*-aristolochene synthase (Starks et al 1997). For this enzyme several amino acids involved in the reaction mechanism have been identified. However, there is not a general model for the large and diverse group of plant terpene synthases.

Boland: A couple of years ago, it was common in organic chemistry to take terpenoid precursors such as farnesol and put them into superacids. Then a lot of cyclization reactions occur spontaneously. It would be interesting to compare the product spectrum of these multiproduct enzymes with that of those random cyclizations: has anyone done this?

Gershenzon: Arigoni did some of that a number of years ago, looking at chemical models for terpene cyclization. What emerges is a clear role for the catalyst. You can get a certain amount of cyclization, for instance, the formation of simple nerolidol or farnesol derivatives from non-enzymatic processes. However, there is no evidence for the formation of complex cyclic sesquiterpenes, such as you get from enzyme-catalysed processes. In answer to the first question, I think that the initial results of the site-directed mutagenesis of terpene synthases show that there's a clear role for the enzyme itself in influencing the product distribution. It does far more than just ionize the diphosphate group to generate an initial carbocation. There is clear participation of the enzyme in other aspects of the reaction.

One of the interesting recent bits of work in this area was an attempt by Joe Chappell and his colleagues to do some domain swapping (Back & Chappell 1996). They reasoned that different domains in the enzyme might be responsible for different parts of the cyclization process. While that idea was a little naïve, what they were able to do in these domain-swapping experiments was to alter the product distribution. And looking at those experiments against the background of the crystal structure, they can begin to infer the participation of particular amino acid residues in catalysis. This work is really just beginning.

Pickett: What about the deprotonation of the pinene carbocation: is that a spontaneous chemical occurrence after the cyclization?

Gershenzon: No, it appears to be an integral part of the cyclization. On the basis of the X-ray structures, there are candidate bases which could participate in the deprotonation process. In considering the overall mechanism, you really couldn't release the final carbocation without the risk of having it quenched by water or another nucleophile. You get an olefinic product, so you infer that it's all being done in a hydrophobic environment.

Gierl: This multiproduct synthesis is almost like combinatorial chemistry. What is the substrate specificity of these multiproduct synthases?

Bohlmann: Monoterpene synthases are highly substrate specific. The same is true for diterpene synthases. Monoterpene and diterpene synthases occur in plastids, which explains the requirement for substrate specificity in order to allow the formation of specific products. Plant sesquiterpene synthases are located in the cytosol and do not only accept FDP but also GDP as substrate *in vitro*. However, this lack of substrate specificity of sesquiterpene synthases *in vitro* does not seem to be relevant *in vivo*, because GDP is produced in plastids.

Pickett: Back in the plastid, does the monoterpene synthase accept neryl pyrophosphate, i.e. the *cis*-isomer of geranyl pyrophosphate?

Gershenzon: Yes, but this is not the native substrate of the monoterpene synthases.

Beale: Isn't there a stereoelectronic requirement for six-member ring formation to go through the tertiary allylic diphosphate, in order to allow 2,3 bond rotation, for the cyclization to occur?

Gershenzon: Yes, neryl pyrophosphate is a good substrate for monoterpenes. In fact, Leopold Ruzicka and colleagues hypothesized the involvement of a neryl cation in monoterpene cyclization, and later workers, such as Derek Banthorpe, supposed that the initial precursor of monoterpene cyclization was neryl pyrophosphate. The argument is really based on the kinetic data for different substrates, and the sorts of abortive intermediates you get with certain substrate analogues. All of it suggests that neryl pyrophosphate is not the native substrate. Geranyl pyrophosphate is a wonderful substrate for the cyclases, and is the ubiquitous intermediate of the main terpene pathway, so there's no need to posit anything else.

Pickett: I was really trying to see how the early literature relates to this very new work.

Boland: If you have multiple products, did you look at the optical purity of the various products? Are these enantiomeric mixtures or are they optically pure? If they are pure, this would mean that each of these transition states is highly controlled.

Pickett: It would be very interesting to put these compounds on a chiral GC column. There would be no need for standards for this study. All this is necessary is to note whether there is a pair of peaks or just one peak, assuming that the chiral column will separate these enantiomeric pairs.

Bohlmann: We have not evaluated optical purity for all products of the two cloned multiple-product sesquiterpene synthases from grand fir. We have done this for the products of multiple-product and single-product monoterpene synthases, and demonstrated that almost all monoterpene products are optically pure enantiomers.

Boland: In that case, I understand it, because you have a single enzyme. In the other case, where you have 50 products, to get optically pure compounds would require the control of each transition state. This would really tell us that this is a multiproduct enzyme with 50 different subunits, each of which controls the folding of one product: this would be the logical consequence.

Beale: With regard to your phylogenetic relationship, *ent*-copalyl diphosphate synthase and *ent*-kaurene synthase seem quite separate. Does this reflect their importance in plant hormone biosynthesis — that they are not really secondary metabolites?

Bohlmann: They are on an isolated and probably early branch. Interestingly, linalool synthase is also found on a distantly related branch, and this is distant from other monoterpene synthases in angiosperms.

Firn: I'm not sure one can make the distinction between gibberellins as plant hormones and secondary metabolites. Most of the 100+ gibberellins have no obvious roles as hormones.

Beale: Only one or two of these are active hormones.

Firn: One could argue that most of the chemical diversity of the gibberellins is simply a consequence of the broad substrate specificity of some of the enzymes involved in their biosynthesis — just like the many examples in secondary metabolism.

Beale: But the diversity comes later, after these cyclase enzymes.

Schultz: Do we know in the case of induced pathways here whether the fungus or the beetle is the inducer? Or is this is a tandem system?

Bohlmann: Mechanical injury is sufficient to induce terpenoid biosynthesis in grand fir at the level of gene activation and enzyme activity of monoterpene, sesquiterpene and diterpene synthases.

Schultz: Do you actually get identical responses from mechanical injury as you do from insect attack?

Bohlmann: We get similar responses, but injury probably doesn't reflect the full complexity of the fungal/bark beetle complex.

Gershenzon: It's worth adding that in this system, in contrast to some of the others mentioned previously, there is a complete redifferentiation of tissue surrounding the wound site. The long time interval before expression of a number of these gene products may simply be a consequence of the time needed to produce entirely new cells at the wound site. This may therefore be very different from some of the other systems we have been discussing.

Eastop: You mentioned differences between individuals. Has anyone looked at the differences between branches on the same trees?

Bohlmann: At the chemical level, yes. There is variation in conifers between younger/older tissues, and branches/needles/stems.

Eastop: But if you take a branch from one side of the tree and one from the other, looking at the same tissues, is there variation between the branches? It is said that on the same tree, different branches are resistant and susceptible to scale insects, which have many generations on the tree.

Bohlmann: I don't know if anyone has looked at that. I would expect a difference. Some of the differences that we see are simply related to the fact that the branches of these trees do have a life history. Forest trees are not grown under controlled conditions: these trees have been outside for longer than 50 years in many cases.

Gershenzon: Rex Cates in Utah has done some nice experiments with Douglas fir, showing significant variation among branches in terpenes and a variety of other constituents (Gambliel & Cates 1995).

References

Back KW, Chappell J 1996 Identifying functional domains within terpene cyclases using a domain-swapping strategy. Proc Natl Acad Sci USA 93:6841–6845

Gambliel HA, Cates RG 1995 Terpene changes due to maturation and canopy level in Douglas-fir (*Pseudotsuga menziesii*) flush needle oil. Biochem Syst Ecol 23:469–476

Starks CM, Back K, Chappell J, Noel JP 1997 Structural basis for cyclic terpene biosynthesis by tobacco 5-*epi*-aristolochene synthase. Science 277:1815–1820

The hydroxamic acid pathway

Alfons Gierl and Monika Frey

Lehrstuhl für Genetik, Technische Universität München, Lichtenbergstrasse 4, 85747 Garching, Germany

Abstract. An important component of general defence mechanisms of plants are toxic secondary metabolites that function as natural pesticides. The cyclic hydroxamic acids DIBOA (2,4-dihydroxy-1,4-benzoxazin-3-one) and DIMBOA (2,4-dihydroxy-7-methoxy-1,4-benzoxazin-3-one) play an important role in the chemical defence of cereals against pests such as insects and pathogenic fungi and bacteria. Five genes that are clustered on chromosome four are sufficient to encode the enzymes to synthesize DIBOA. The first gene in the pathway, *Bx1*, encodes an enzyme resembling a tryptophan synthase α subunit that catalyses the formation of indole and thereby establishes the branchpoint that leads to the secondary metabolites. Four cytochrome P450-dependent monooxygenases encoded by *Bx2–Bx5* catalyse consecutive hydroxylations to form DIBOA. This pathway can be generalized for grasses, since identical enzyme activities have been found in rye. The pathway is relatively short and begins with a metabolite ubiquitous to plants. Therefore DIBOA biosynthesis could be introduced into other plant species to confer improved disease resistance.

1999 Insect–plant interactions and induced plant defence. Wiley, Chichester (Novartis Foundation Symposium 223) p 150–159

A substantial number of secondary metabolites in plants are dedicated to pathogen defence. These include the cyclic hydroxamic acids DIBOA (2,4-dihydroxy-4-benzoxazin-3-one) and its methoxy derivative DIMBOA (2,4-dihydroxy-7-methoxy-1,4-benzoxazin-3-one) which are found almost exclusively in Gramineae. DIMBOA confers resistance to first brood European corn borer (*Ostrinia nubilalis*), northern corn leaf blight (*Helminthosporium turcicum*), maize plant louse (*Rhopalosiphum maydis*) and stalk rot (*Diplodia maydis*), as well as to the herbicide atrazine (Niemeyer 1988). The DIBOA and tryptophan biosynthetic pathways share certain intermediates. Labelled tryptophan precursors such as anthranilic acid and indole are incorporated into DIMBOA, although labelled tryptophan is not incorporated (Frey et al 1997).

In maize, a series of five genes that are clustered on the short arm of chromosome four is sufficient to encode the enzymes for DIBOA synthesis (Frey et al 1997). The first gene in the pathway, *Bx1*, encodes an enzyme resembling a tryptophan synthase α subunit (Frey et al 1997, Melanson et al 1997) that catalyses the formation of free indole (Frey et al 1997) and thereby establishes the branchpoint that leads to the secondary metabolites DIBOA and DIMBOA. The next four genes, *Bx2–Bx5*, encode cytochrome P450-dependent monooxygenases of the CYP71C subfamily (Frey et al 1995), that catalyse four hydroxylations including one ring-expansion to form the defence compound DIBOA (Fig. 1). The classic forms of cytochrome P450 monooxygenases are membrane-bound haem-containing mixed function oxidases. They utilize NADPH to reductively cleave

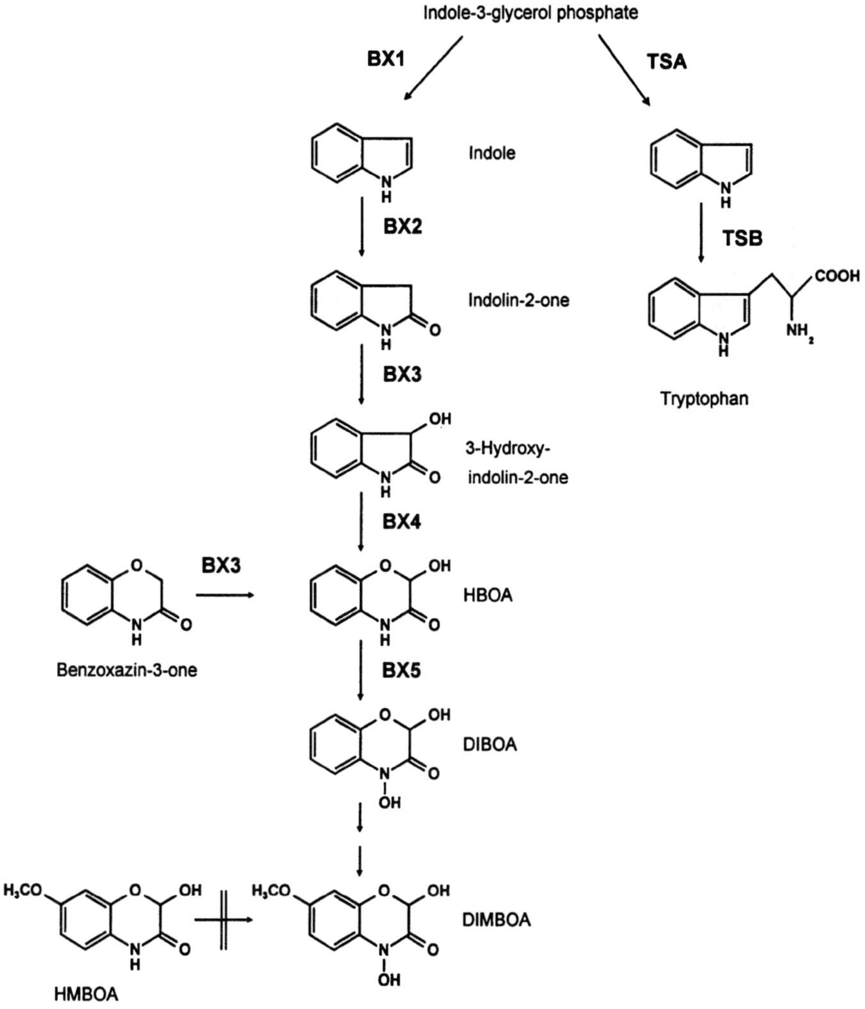

FIG. 1. DIMBOA biosynthetic pathway. The branch point from tryptophan biosynthesis is shown on top. The DIMBOA-specific enzymes are indicated by BX1–BX5.

molecular oxygen to produce functionalized organic products and a molecule of water. In this generalized reaction, reducing equivalents from NADPH are transferred to the P450 enzyme via a flavin-containing NADPH-P450 reductase (CPR). In plants, P450 enzymes are involved mainly in hydroxylation or oxidative demethylation reactions (Donaldson & Luster 1991) of a large variety of primary and secondary metabolites including hormones, phytoalexins and pharmaceutically relevant compounds. These reactions can also result in detoxification of xenobiotics.

Isolation and functional analysis of the *Bx1* gene

The maize mutation *bx1* (*benzoxazinless*) abolishes DIMBOA synthesis (Hamilton 1964). Plants homozygous for *bx1* grow normally but are extremely susceptible to the above mentioned pathogens. The *Bx1* gene was cloned by directed transposon tagging using the *Mutator (Mu)* transposon system of maize (Chomet 1994). A *Mu*-induced recessive *bx1* allele was identified. Subsequently, a genomic DNA fragment flanking the *Mu* insert was isolated by the so-called AIMS method (a PCR-based method, Frey et al 1998). This fragment was used to isolate the wild-type *Bx1* and the recessive *bx1* (Frey et al 1997) alleles from genomic λ libraries as well as a full length cDNA clone. DNA sequence analysis revealed the structures of these three alleles. Another allele of *Bx1* was accidentally isolated by searching for genes that are not expressed in the maize cob by Kramer & Koziel (1995). *Bx1* has similarity to tryptophan synthase α subunits (TSA) and when expressed in *Escherichia coli* complements a bacterial TSA mutation. TSA catalyses the conversion of indole-3-glycerol phosphate to indole, the penultimate reaction in tryptophan biosynthesis. It was shown previously that indole is an intermediate in DIMBOA biosynthesis (Desai et al 1996). It was concluded that the *bx1* mutant should be defective in the production of free indole.

We therefore complemented *bx1* with indole. The immersion of shoots of *bx1* seedlings (4 days after imbibition) into a 1 mM solution of indole in the dark for one day restored the formation of DIMBOA. This confirmed that indole is the first specific intermediate for DIMBOA biosynthesis after branching from primary metabolism.

In bacteria, TSA activity is almost completely dependent on formation of a complex with tryptophan synthase β (TSB), and indole is usually not released during tryptophan synthesis. An analogous heterosubunit complex exists in *Arabidopsis* (Radwanski et al 1995). If the TSA homologue BX1 catalyses the formation of free indole from indole-3-glycerol phosphate, BX1 should function independently of TSB. This assumption was tested by expression of *Bx1* in *E. coli* and purification of the BX1 protein to homogeneity. The steady-state kinetic constants were measured. The Michaelis constant K_m was determined to

0.013 mM (indole-3-glycerol phosphate) and the maximum velocity k_{cat} to 2.8 s^{-1}. Comparison of these values with the constants for indole-3-glycerol phosphate cleavage of *E. coli* tryptophan synthase complex ($K_m = 0.027$ mM; $K_{cat} = 0.2$ s^{-1}) demonstrated that BX1, independent of TSB, is approximately 30 times more efficiently catalysing the production of indole compared to the bacterial complex. Therefore, BX1 acts efficiently as an indole-3-glycerol phosphate lyase to produce indole. Since tryptophan levels in the *bx1* mutants are normal, the natural TSA subunit must be encoded by gene(s) different from *Bx1*. The product of these genes would form the heterosubunit complexes that are typical for tryptophan synthases. *Bx1* is required for the production of free indole and secondary DIMBOA synthesis.

The BX1 protein, like tryptophan synthase, is localized in the stroma of the chloroplast. This was indicated by import experiments with spinach chloroplasts. The nature of the BX signal peptide was deduced from the determination of the N-terminal amino acid sequence of the mature protein (C. Stettner & M. Frey, unpublished results).

The synthesis of several other secondary metabolites in plants, such as the indole glucosinates, anthranilate-derived alkaloids and tryptamine derivatives (Radwanski & Last 1995, Kutchan 1995), depends on the tryptophan pathway. Indole-3-glycerol phosphate was also proposed as a branchpoint from the tryptophan pathway for the synthesis of the indolic phytoalexin camalexin (3-thiazol-2'yl-indole) in *Arabidopsis thaliana* (Tsuji et al 1993, Zhao & Last 1996). It will be interesting to see whether a BX1 homologous enzyme would also catalyse the first specific step in camalexin synthesis, indicating a mechanism by which indole-3-glycerol phosphate could serve as an intermediate in a wide range of secondary metabolites in plants.

Isolation and functional analysis of the *Bx2–Bx5* genes

Four maize cytochrome P450-dependent monooxygenase genes, one of which was isolated by subtractive cDNA cloning from high versus low DIMBOA accumulating lines (Frey et al 1995), are members of the *CYP71C* subfamily of plant cytochrome P450 genes. These genes are strongly expressed in young maize seedlings, share an overall amino acid identity of 45–60%, and are clustered on the short arm of chromosome 4. The finding that all oxygen atoms of DIMBOA are incorporated from molecular oxygen (Glawischnig et al 1997), led to the speculation that these cytochrome P450 enzymes might be involved in this pathway. The genes encoding these enzymes are designated *Bx2*, *Bx3*, *Bx4* and *Bx5* in the rest of this chapter.

Direct evidence for the involvement of *Bx3* in DIMBOA biosynthesis is provided by a mutant allele (*Bx3::Mu*) isolated by a reverse genetic approach to screen for *Mu* insertions in the P450 genes (Frey et al 1997). In maize seedlings homozygous for the recessive mutant allele, no DIMBOA could be detected by HPLC analysis. In contrast, DIMBOA was detected in seedlings that were either heterozygous or homozygous wild-type. Thus an intact *Bx3* gene is required for DIMBOA biosynthesis.

In order to demonstrate the function of the four P450 enzymes in DIMBOA biosynthesis, we used a yeast expression system (Truan et al 1993). The cDNAs of *Bx2–Bx5* were inserted into the pYeDP60 expression vector (Urban et al 1994). These constructs were used to transform the WAT11 yeast strain. In WAT11, a galactose-inducible *Arabidopsis thaliana* microsomal NADPH-P450 reductase (*ATR1*) replaces the yeast reductase (Pompon et al 1996).

Microsomes were isolated from the transgenic yeast strains and tested for enzymatic activity (Truan et al 1993, Pompon et al 1996). BX2–BX5 convert indole to DIBOA by catalysing a series of four specific hydroxylations (Fig. 1). This includes the conversion of 3-hydroxy-indolin-2-one to 2-hydroxy-1,4-benzoxazin-3-one. The mechanism for this unusual ring expansion is currently being analysed.

Bx2–Bx5 have probably evolved by gene duplication events as indicated by clustering of the genes on the short arm of chromosome four and the sequence similarity between these genes. They have been grouped into the CYP71C subfamily (Frey et al 1995) of cytochrome P450-dependent monooxygenases.

Although the four cytochrome P450 enzymes are homologous proteins, they are substrate specific. Only one substrate was converted by each respective P450 enzyme to a specific product. The question of enzyme specificity was addressed with the yeast expression system. The intermediate metabolites of the DIBOA pathway indole, indolin-2-one, 3-hydroxy-indolin-2-one, and 2-hydroxy-1,4-benzoxazin-3-one (HBOA) (Frey et al 1997) were incubated with microsomal preparations each containing one of the P450 enzymes. No detectable conversions occurred in other enzyme/substrate combinations. Each enzyme is therefore specific for the introduction of only one of the oxygen atoms of the DIBOA molecule. Enzymatic reactions, identical to the ones with the different yeast microsomal preparations, could be performed with maize microsomes, indicating that these reactions occur natively in maize. These findings suggest an *in vivo* reaction sequence in maize from indole to DIBOA.

In addition to the reactions described above, benzoxazin-3-one and 2-hydroxy-7-methoxy-1,4-benzoxazin-3-one (HMBOA, Fig. 1) were tested as possible substrates for BX2–BX5. The hydroxylation of benzoxazin-3-one to HBOA is catalysed by BX3 (Glawischnig et al 1999). Hydroxylation of benzoxazin-3-one is approximately half as efficient as the hydroxylation of indolin-2-one. In both cases a C atom at an equivalent position is hydroxylated (Fig. 1). In contrast, for HMBOA, a derivative present in maize seedlings, no enzymatic conversions are detectable with BX2–BX5 expressed in yeast or with maize microsomes. In particular, HMBOA is not N-hydroxylated by BX5. This indicates that the 7-methoxy group of HMBOA would interfere with BX5 action.

The relatively high specificity of the enzymes seems to support the idea that plant P450s generally have a much greater substrate specificity than their animal homologues. However, there is emerging evidence that plant P450s in addition to their normal physiological function, can also convert certain xenobiotics with varying efficiencies.

DIBOA biosynthetic enzymes in other cereals

The hydroxamic acids are widely distributed in grasses (Niemeyer 1988), suggesting that the acquisition of this pathway occurred relatively early in the evolution of the Gramineae. The activity of the DIBOA-specific P450 enzymes was assayed in two other cereals, rye (containing hydroxamic acids) and barley (without hydroxamic acids). These two species are much more closely related to each other than either is to maize (Devos & Gale 1997).

The major cyclic hydroxamic acid of rye is DIBOA. As in maize, there are relatively high concentrations (up to 1 mg/g fresh weight) present in rye seedlings. In order to investigate whether identical cytochrome P450-dependent reactions are associated with DIBOA biosynthesis in rye, we isolated microsomes from seedlings. The CPR activity in rye microsomes is about 1.8-fold higher than in maize. All four substrates tested were converted to products that were identical to the products obtained with maize microsomes (Glawischnig et al 1999). No additional products were detected. The reactions were strictly dependent on NADPH indicating true cytochrome P450 enzyme reactions. The specific activities determined for the rye microsomes were very similar to the values detected with maize microsomes for the first three reactions in the pathway. The conversion of HBOA to DIBOA was reproducibly lower (a factor of two) in rye microsomes. The identity of the reactions in maize and in rye suggests identical DIBOA biosynthetic pathways for both species.

In microsomes prepared from barley seedlings, no activities of the P450 enzymes of the DIBOA pathway were detectable, although a total P450 content of 3 μg/mg microsomal protein and a significant CPR activity was determined (Table 1).

Maize and rye are two distant species among the Gramineae. However, the DIBOA biosynthetic pathway is identical in both species. Therefore, a set of proteins homologous to BX2–BX5 can be proposed in other grasses. If this were the case, the duplications responsible for the evolution of the *Bx2–Bx5* gene cluster must have occurred early in the development of the Gramineae. In maize, the *Bx1* gene is included in this cluster. The isolation of genes homologues to *Bx1–Bx5*

TABLE 1   DIBOA biosynthetic P450 enzyme activities in maize, rye and barley

	Specific activity of microsomes ($nmol \times mg^{-1} \times min^{-1}$) with the substrate				
Plant	Indole	Indolin-2-one	Benzoxazin-3-one	3-Hydroxy-indolin-2-one	HBOA
Maize	0.3	0.24	0.09	0.94	0.86
Rye	0.25	0.19	0.07	0.76	0.45
Barley	—	—	—	—	—

from rye and other grasses could give insight into the evolution of the *Bx* gene cluster.

Barley microsomes show no significant DIBOA-specific P450 activities. It remains to be shown whether the loss of enzyme activity is due to gene inactivation, or whether the whole *Bx2–Bx5* cluster has been lost. This might have occurred during agricultural breeding from wild barley varieties in which DIBOA was present (Niemeyer 1988). A similar loss of enzyme activity has been observed for the UDP-glucose:DIBOA glycosyltransferase. Glycosylation is required for transport of DIBOA into the vacuole. Glycosyltransferase activity present in wild varieties was also lost during barley cultivation (Leighton et al 1994).

References

Chomet P 1994 Transposon tagging with mutator. In: Freeling M, Walbot V (eds) The maize handbook. Springer-Verlag, New York, p 243–249

Desai SR, Kumar P, Chilton WS 1996 Indole is an intermediate in the biosynthesis of cyclic hydroxamic acids in maize. Chem Commun 11:1321

Devos KM, Gale MD 1997 Comparative genetics in the grasses. Plant Mol Biol 35:3–15

Donaldson RP, Luster DG 1991 Multiple forms of plant cytochrome P_{450}. Plant Physiol 96:669–674

Frey M, Kliem R, Saedler H, Gierl A 1995 Expression of a cytochrome P_{450} gene family in maize. Mol Gen Genet 246:100–109

Frey M, Chomet P, Glawischnig E et al 1997 Analysis of a chemical plant defense mechanism in grasses. Science 277:696–699

Frey M, Stettner C, Gierl A 1998 A general method for gene isolation in tagging approaches: amplification of insertion mutagenized sites (AIMS). Plant J 13:717–721

Glawischnig E, Eisenreich W, Bacher A, Frey M, Gierl A 1997 Biosynthetic origin of oxygen atoms in DIMBOA from maize: NMR studies with $^{18}O_2$. Phytochemistry 45:715–718

Glawischnig E, Grün S, Frey M, Gierl A 1999 Cytochrome P_{450} monooxygenases of DIBOA biosynthesis: specificity and conservation among grasses. Phytochemistry 50:925–930

Hamilton RH 1964 A corn mutant deficient in 2,4-dihydroxy-7-methoxy-1,4-benzoxazin-3-one with an altered tolerance of atrazine. Weeds 12:27–30

Kramer VC, Koziel MG 1995 Structure of a maize tryptophan synthase alpha subunit gene with pith enhanced expression. Plant Mol Biol 27:1183–1188

Kutchan TM 1995 Alkaloid biosynthesis—the basis for metabolic engineering of medicinal plants. Plant Cell 7:1059–1070

Leighton V, Niemeyer HM, Jonsson LMV 1994 Substrate specificity of a glucosyltransferase and an N-hydroxylase involved in the biosynthesis of cyclic hydroxamic acids in *Graminea*. Phytochemistry 36:887–892

Melanson D, Chilton M-D, Masters-Moore D, Chilton WA 1997 Deletion in an indole synthase gene is responsible for the DIMBOA-deficient phenotype of *bxbx* maize. Proc Natl Acad Sci USA 94:13345–13350

Niemeyer HM 1988 Hydroxamic acids (4-hydroxy-1,4-benzoxazin-3-ones), defence chemicals in the Gramineae. Phytochemistry 27:3349–3358

Pompon D, Louerat B, Bronine A, Urban P 1996 Yeast expression of animal and plant P_{450}s in optimized redox environments. Methods Enzymol 272:51–64

Radwanski ER, Last RL 1995 Tryptophan biosynthesis and metabolism: biochemical and molecular genetics. Plant Cell 7:921–934

Radwanski ER, Zhao J, Last RL 1995 *Arabidopsis thaliana* tryptophan synthase alpha: gene cloning, expression, and subunit interaction. Mol Gen Genet 248:657–667

Truan G, Cullin C, Reisdorf P, Urban P, Pompon D 1993 Enhanced *in vivo* monooxygenase activities of mammalian P_{450}s in engineered yeast cells producing high levels of NADPH-P_{450} reductase and human cytochrome b_5. Gene 125:49–55

Tsuji J, Zook M, Hammerschmidt R, Last RL, Sommerville SC 1993 Evidence that tryptophan is not a direct biosynthetic intermediate of camalexin in *Arabidopsis thaliana*. Physiol Mol Plant Pathol 43:221–229

Urban P, Werck-Reichhart D, Teutsch HG et al 1994 Characterization of recombinant plant cinnamate 4-hydroxylase produced in yeast. Kinetic and spectral properties of the major plant P_{450} of the phenylpropanoid pathway. Euro J Biochem 222:843–850

Zhao J, Last RL 1996 Coordinate regulation of the tryptophan biosynthetic pathway and indolic phytoalexin accumulation in *Arabidopsis*. Plant Cell 8:2235–2244

DISCUSSION

Bohlmann: With respect to the regulation of tryptophan and DIMBOA biosynthesis, you are saying that under stress conditions, the plant produces much more DIMBOA indole-derived metabolites than tryptophan. We know that tryptophan biosynthesis is tightly feedback-inhibited at the site of anthranilate synthase. Therefore, one has to postulate independent precursor supply for tryptophan and DIMBOA biosynthesis. Are these two pathways differentially compartmentalized?

Gierl: This is exactly what we would like to know. First of all, the end product of DIMBOA biosynthesis is stored in the vacuole and so it might not be relevant for feedback inhibition of anthranilate synthase. Since many of these metabolic genes in the primary pathway are duplicated genes, they may act in special compartments: one compartment for primary and one for secondary metabolism. We now have the tools to look at where these enzymes are expressed at the subcellular level.

Dietrich: Did you say that rice does not produced DIMBOA?

Gierl: Yes.

Dietrich: Do you know from the rice genomic data whether the corresponding genes are present in rice?

Gierl: We are initiating a collaboration with a Korean group that is part of the rice genome project. Rice may be similar to barley in that the wild varieties may have DIMBOA. As you probably know, the genome of maize was duplicated about 25 million years ago, so all maize genes are present in duplicated form although nowadays they usually have a specialized function. However, this segment of chromosome 4 short arm does not belong to this duplication. We have made an attempt using hybridization to get the genes from wheat in order to see whether their DIMBOA gene cluster is identical, but maize and wheat are relatively distant, so that hybridization did not work and we will have to use PCR.

Pickett: Are these genes expressed in the seed or on the tissue that forms the seed?

Gierl: No, they are induced very early after germination. The highest expression levels occur about 5–7 days after germination.

Pickett: I am interested in the development of any of these genes in crop protection. In this direction, is it possible to keep the gene products out of the seed so that this is safe for human consumption?

Gierl: This is what the breeders actually managed in maize. After 4 or 6 weeks of development there is almost no activity detectable and DIMBOA levels in the seeds are very low. Since these compounds are toxic, I would like to have them first expressed in non-food transgenic plants — cotton would be very nice. Interestingly, many plants, including cotton, have the potential to produce indole, the precursor of DIMBOA, so perhaps in such plants we could rely on relatively large substrate pools for the production of DIMBOA.

Pickett: I don't think there's any real problem with them being in food plants if they are not expressed directly in the seed. Alkaloidal components of potatoes can be very toxic but, being largely confined to the leaves, we eat the tubers happily. Barley is cyanogenic, and apart from a recent scare from the Food and Drugs Administration in the USA, where for a time it appeared that they were going to ban scotch whisky imports, there is also no problem.

Gierl: I agree, but the picture is simply different if one wants to market a transgenic plant expressing a toxic compound.

Firn: How well do mutant maize plants that can't make these compounds resist insect and fungal attack?

Gierl: The *bx1* mutants which lack all these compounds are heavily infested by insects and fungi. The problem with respect to the European corn borer that I have shown is that maize during the juvenile development is highly resistant. However, when the pathway is shut down and the DIMBOA concentrations drop, it is infested.

Scheel: Does the area where you find the *Bx2–Bx5* cluster contain other P450-encoding genes?

Gierl: Paul Sisco (North Carolina State University) has isolated a mutant for the P450 enzyme that is required for the OCH$_3$ group of DIMBOA. This mutant maps close to the *Bx* gene cluster. The methyltransferase also maps to this region of chromosome 4. I haven't got a good explanation for this. Perhaps this cluster is more easily transmitted as a whole, which may be advantageous, or it may simply reflect an ancient evolutionary event.

Mitchell-Olds: With the lines that don't express these genes, is this due to mutations or natural variation in enzyme-encoding loci?

Gierl: In the case of the *bx1* mutant, it is deletion of the promoter region. In the low-expressing lines we don't know what has happened.

Gershenzon: Is there anything known about other plant systems in which there is this clustering of sequential genes in a biosynthetic pathway?

Gierl: With respect to plants, the clustering of genes from one pathway is an unusual situation. In the case of fungi, this is commonly found in the polyketide synthases for antibiotics.

Gershenzon: You proposed an attractive evolutionary scenario in which there is a gene duplication followed by evolution of a new enzyme that is able to catalyse an additional reaction in an expanding biosynthetic sequence. Is there any evidence that the particular intermediates are themselves toxic to the plant? In other words, in thinking about the selection pressures that may have led to the evolution of this plant pathway, perhaps the cytotoxicity of a metabolite selected for an enzyme that could further metabolize it to avoid negative consequences.

Gierl: The intermediates are relatively stable. The molecule begins to become reactive by gaining this NOH group, which is one of the last steps in the biosynthesis. Then, during decomposition, there are again some reactive species. The decomposition products in particular also have a role as allelopathic compounds, inhibiting germination of other plants.

Scheel: You mentioned that DIMBOA is stored as a conjugate in the vacuole. At which step does the conjugation occur?

Gierl: We are trying to clone the enzyme. DIMBOA is glycosylated at the 2 position. Logically, this would be the last step.

Boland: I have a mechanistic question. The biosynthesis has one particularly interesting step: the C–C cleavage with the O insertion. Is the mechanism behind this known?

Gierl: I'm not a chemist, but we have collaborated with a chemist about that, and there are actually two possibilities. Either the oxygen of the 2 position is retained in the heterocycle of the benzoxazinone, or the initial reaction is the epoxidation of the aromate. Our experiments indicate that the latter is the case.

Boland: That is what I would expect.

General discussion II

Volicitin and the induced volatile response

Firn: Because volicitin is not something which a herbivore has to make, and there seems to be a great disadvantage to any insects making it in terms of inducing the plant volatiles, hence attracting parasites or predators, one would imagine that there is quite a strong selective pressure against volicitin production. One wonders why herbivores still produce it. If there is no adequate explanation for this, one would have to adopt the radical view that the plant volatile production may be advantageous to the herbivores.

Pickett: It could be that volicitin does have a function for the caterpillar as a salivary component, or as a pheromone employed in spacing behaviour by the herbivore.

Lamb: The obvious model would be that the caterpillar produces this to block the induction of the direct defences. The plant has then responded in a vicious way by using it as a signal for the indirect defences.

Pickett: Jim Tumlinson, have you any evidence of volicitin blocking direct defences?

Tumlinson: No; that's another one of the experiments that needs to be done. Volicitin probably turns on the octadecanoid signalling pathway, which is involved in the production of proteinase inhibitors and other compounds like that, which would be direct defences.

Lamb: However, one of the new messages for me this morning was that there's a repertoire of different octadecanoid signals; they don't necessarily overlap terribly well.

Tumlinson: That's true; we don't know the function of all the components.

Turlings: An interesting observation during the isolation process of volicitin was that that each time we got it cleaner it was more active. This suggests that there are also other compounds in the regurgitant that suppressed its activity.

Pickett: This in turn suggests that the caterpillar may have evolved a mechanism that reduces the effect of the volicitin.

Lamb: Did you ever mix back components of the spit? Does one component affect another?

Tumlinson: We have identified all these compounds and synthesized them, and now we are seeing if we can put them back together. We haven't completed that experiment, but so far the only compounds that we can attribute any volatile-inducing activity to are volicitin and the analogue without a hydroxyl group.

Vet: Does volicitin alone have the same degree of inducing effect as the whole spit complex together?

Tumlinson: Yes, it does. One of the first things we did was to compare the synthetic compound with crude spit; it has a similar activity. One of the problems here is that of using the corn seedling bioassay in which we are cutting off the corn seedling and letting it take up the material through the petiole. We haven't done this with other plants, so there may be a different activity of different components on different plants. The other problem with this bioassay is that the caterpillar doesn't do it this way. The caterpillar chews this material into the plant over a long period.

Boland: One crucial experiment to clarify whether volicitin or the spit do have different inducing activity would be to look at the late volatiles. One of the crucial points is the presence of *cis*-jasmone in volatile blends. In the cotton you observed *cis*-jasmone. If volicitin does not induce the octadecanoid pathway, the question arises as to where the *cis*-jasmone comes from. This would answer the question as to whether volicitin is the only competent component in the secretion.

Poppy: Getting back to the selective driving forces behind this, I think Richard Firn does have a point here: if volicitin causes such a negative impact on the caterpillars, why is it still there? We can draw an analogy here with the haematophagous insects. Many of these respond to CO_2, which is a smart move, because it is hard for evolution to select animals that don't breathe out CO_2.

Pickett: What are the properties of volicitin? Is it a surfactant?

Tumlinson: We haven't made it in very large quantities, but it should be a good surfactant.

Schultz: I'd be curious about how good a detergent volicitin is, because insects don't have gall bladders. As a consequence, in order to create the critical micelle levels for them to be able to absorb fatty acids, they use lipids: fatty acids themselves, phospholipids, lysophospholipid. The surface tension of caterpillar guts is nothing like water: it is strongly detergent. Several of us have demonstrated that it is plant-acquired and processed fatty acids and lipids that provide this property. If insects lack this detergent gut, they can't acquire fatty acids from the diet. Thus the digestive tracts of all insects are packed with detergent lipids. It could be that volicitin is serving this purpose.

Firn: The structural requirements for detergent-like activity are less demanding than the requirements for specific, potent biological activity. Hence, if a detergent-like activity were beneficial, one would have expected the insect to have evolved such a substance which had no disadvantageous secondary biological activity.

Schultz: Except that there does appear to be the constraint in insects that they can't use bile salts the way vertebrates do. In insects detergents are always fatty acid based.

Tumlinson: It appears that the plants which are normally attacked by a certain species of insect respond more strongly (i.e. release larger quantities of volatiles) to that species of insect than to an insect that doesn't normally attack the plant. In some of the work that Consuelo De Moraes did, where she showed that the wasp was picking *Heliothis virescens* over *Helicoverpa zea*, it turns out that tobacco does respond more strongly to *H. virescens* than it does to *H. zea*, and corn appears to respond more strongly to *H. zea* than to *H. virescens*. This relationship would suggest that the plant is responding to the herbivore rather than the other way round.

Vet: Even if the caterpillars have volicitin for a particular function, can't there be selection on them not to release it onto the leaf surface during feeding?

Tumlinson: That brings up a completely new set of questions. We have obtained these compounds by squeezing the caterpillar and causing it to regurgitate. The caterpillar doesn't necessarily regurgitate on the plant as it chews, so we may be introducing other artefacts there also — it's just that it's very difficult to get this material in any other way.

Dicke: Aphids do salivate into the plant.

Poppy: But the aphid situation is quite different. According to many researchers, there are two types of saliva, one which gels around the stylets as they penetrate the plant, and another, the watery saliva, which is supposed to be more involved in the actual feeding process, but probably only entering the phloem to initiate feeding, thereafter only facilitating feeding in the aphid.

Scheel: An easy way to avoid the recognition of volicitin by the herbivore would be to use another amino acid. Is there a good reason why glutamine hasn't been used?

Tumlinson: But then the plant would evolve to respond to this new amino acid. The herbivores that normally feed on the plant elicit a stronger response from that plant than the ones that don't. I don't know why this is, but there's some sort of close relationship there.

Pickett: What about the other insects you're looking at, Jim Tumlinson? Are you able to say anything to say about other elicitors?

Tumlinson: We are looking at the hornworm, as is Ian Baldwin. The crude hornworm spit is not as active on corn seedlings as the beet armyworm spit. We don't yet have all the compounds isolated and we're not really sure what's going on, because our other bioassays don't work too well yet. Ted Turlings showed a long time ago that grasshopper spit is 5–10 times as active on corn seedlings as any Lepidopteran spit. We have isolated and purified the elicitor responsible, and all I can say is that it is not exactly like volicitin but we don't yet know the complete structure.

Pickett: Is that a specialist feeder?

Tumlinson: Not really. I have no idea why its spit is more active.

Schultz: We heard yesterday that some people are coming up with insects that don't elicit direct induced responses. I have one; Gary Felton is arguing that glucose oxidase in saliva of some insects he is studying shuts down some of these induced responses, but I don't know anybody who has a good example of that with the volatile system.

Tumlinson: There must be insects that don't induce volatiles.

Dicke: When whiteflies infest cucumber plants, their parasitoids do not respond to any volatiles from the plants (Noldus & van Lenteren 1990). However, it is not clear whether this is the failure of the parasitoids to respond to volatiles or whether the plant simply doesn't produce them.

Tumlinson: This summer we looked at beet armyworms and we found a great variability in the quantities and relative proportions of the constituents of spit: some of them had very little volicitin. We don't know whether that's just an artefact of our rearing or handling procedures and so forth.

Vet: Are you going to select lines to study this variability?

Tumlinson: At this point we're just trying to figure out what's going on.

Pickett: For *Cotesia sesamiae* foraging for its noctuid host, it seems to be using mainly the nonatriene that we have discussed here repeatedly: this compound accounts for a large amount of activity from the damaged plant, in this particular case (Khan et al 1997). Usually with herbivores, 20 components of a 300–500 compound mixture are involved in the interaction, and at a very wide range of concentrations. The system is complex, but only certain aspects are involved and usually a limited range of compounds, with the number being similar, for a whole range of insect–host interactions.

The plant is resourcing production of a wide range of compounds, but many are not used in these recognition process, the implication being that they may be there for some other reason. On the surface of plants there are numerous spores of pathogens that will develop opportunistically when a lesion is formed. This rapid burst of volatiles associated with such a lesion might impede the initial development of these spores. We may also need to investigate the minor components for this effect if the major ones are involved.

Gershenzon: I'm struck by the fact that although the synthesis of herbivore-induced volatiles is induced *de novo*, as Paul Paré and Jim Tumlinson have shown (Paré & Tumlinson 1997), there always seems to be a bit of a delay between the initiation of herbivore damage and the production of these compounds. If these substances are so important as attractants for parasitoids and predators of the higher trophic levels, why do plants risk this delay? We know in a number of cases that plants can release metabolites rapidly from hydrolyzable conjugates, for example, or from storage reservoirs. Either way is faster than starting up the machinery of isoprenoid biosynthesis from scratch. In this context, I was really interested in Marcel Dicke's comment yesterday, that you can actually just shake the plants and get a release of volatiles. Here there is no delay at all.

Lamb: I assume that the plant doesn't want to send the wasp misleading information. It would have to be inducible.

Dicke: It's not necessarily misleading information if it is sent out immediately on attack and its emission stops when attack ends.

Lamb: It might be that the plant has to go through a series of checkpoints. Otherwise, how would it distinguish grazing from a gust of wind and so on?

Turlings: The response of corn plants that we observe is very quick — within hours. If we assume that it starts off with a tiny little caterpillar, this caterpillar will hardly do any damage at all to the plant in that time.

Yesterday we were talking about this possible trade-off between indirect and direct defences. In cotton we see a much longer delay in the production of these induced volatiles, and it seems that cotton has constitutive defences that are immediately operative, and the plant may decide to rely on those constitutive defences until it finds out whether or not they are working before sending these volatile signals.

Boland: We haven't discussed the emission of ethylene, which is a faster reaction: something which happens perhaps after one hour. Furthermore, we have an immediate emission of the green leaf volatiles. What about other lipoxygenase products such as pentane, hexane and ethane? These haven't yet been excluded as possible signalling components.

Pickett: We have a highly volatile compound emitted very early, which we believe relates directly to the oxidative burst.

Tumlinson: Could it be a matter of investment? Until there's a significant amount of damage, perhaps it isn't worth the plant investing in starting off all of these biosynthetic processes.

Pickett: The wasp may wait for longer-term and more useful stress-related semiochemical information.

Lamb: This was my point: if the wasp comes and there's a tiny, grotty little caterpillar, it is not a very useful signal for it.

Vet: The parasitoids do show a very strong density-dependent response: they don't go for a plant with one little caterpillar.

Lamb: So where the plant is responding to a heavy infestation, it is a more meaningful signal.

Vet: On the other hand, if the plant uses a specialized parasitoid like *Cotesia marginiventris* which only attacks young stages, then of course they should not wait too long.

Schultz: I feel obligated to point out that although this is a fun conversation, there may be nothing whatsoever adaptively significant about the timing. Among other things, insects don't matter that much to the plants, I'm sorry to say. If this were a pathogen, then the plant would need to stop it in the next 12 hours. If you want to make this a strictly adaptive argument I would argue that there isn't that much selection on the plant to improve its performance, but it's also worth considering that adaptation may have nothing to do with this and the plants are running as fast as they can. Besides, you're talking about say six plant species: is that sufficient to generalize about the rapidity of this process? This conversation has got awfully adaptive, and it makes me nervous.

Boland: When we started homoterpene biosynthesis, we did it with labelled precursors. We went into the botanical garden and took about 70 species of different angiosperms. Out of these 70 tested, 50 were positive, so this is not something which is for specialized plants. We didn't find the ability to produce these compounds in gymnosperms, so homoterpenes appear to be restricted to angiosperms.

Schultz: The ability to do this certainly is widespread, but the speed at which it can be done could be all over the map.

Karban: Any particular herbivore is not necessarily exerting selection or affecting plant fitness, but there are certainly examples of plants that are strongly affected by herbivores. For the last three or four summers I have been doing some work on *Nicotiana attenuata*. Not only can those big hornworms do in a plant completely, but also grasshoppers, noctuids and corimelaenid seedbugs can all have big effects on plant fitness.

Schultz: I wouldn't dispute that. I just get nervous when we work very hard to explain extremely subtle details of plant behaviour on the basis of one taxon.

References

Khan ZR, Ampong-Nyarko K, Chiliswa P et al 1997 Intercropping increases parasitism of pests. Nature 388:631–632

Noldus LPJJ, van Lenteren JC 1990 Host aggregation and parasitoid behaviour: biological control in a closed system. In: Mackauer M, Ehler LE, Roland J (eds) Critical issues in biological control. Intercept, Andover, p 229–262

Paré PW, Tumlinson JH 1997 *De novo* biosynthesis of volatiles induced by insect herbivory in cotton plants. Plant Physiol 114:1161–1167

Cross-talk between the signal pathways for pathogen-induced systemic acquired resistance and grazing-induced insect resistance

G. W. Felton*, J. L. Bi*, M. C. Mathews*, J. B. Murphy*, K. Korth†, S. V. Wesley†, C. Lamb‡¶ and R. A. Dixon†

*Departments of Entomology and Horticulture, University of Arkansas, Fayetteville, AR 72701, USA, †Plant Biology Division, Samuel Roberts Noble Foundation, 2510 Sam Noble Parkway, Ardmore, OK 73401, USA, and ‡University of Edinburgh, Institute of Cell and Molecular Biology, Swann Building, Mayfield Road, Edinburgh EH9 3JR, UK

Abstract. Reducing phenylpropanoid biosynthesis in transgenic tobacco compromises systemic acquired resistance (SAR) to tobacco mosaic virus, while increasing phenylpropanoid biosynthesis enhances SAR. Surprisingly, transgenic tobacco plants compromised in SAR exhibit more effective grazing-induced systemic resistance to larvae of Heliothis virescens, whereas induced insect resistance is compromised in transgenic plants with elevated phenylpropanoid levels. Levels of the phenylpropanoid-derived signal salicylic acid are directly correlated with overall phenylpropanoid biosynthesis in this series of transgenic plants. Moreover, while pathogen-induced SAR is almost completely compromised in salicylic acid-deficient plants expressing the bacterial nahG salicylate hydroxylase gene, these plants show enhanced grazing-induced insect resistance compared to wild-type. Hence, suppression of grazing-induced insect resistance is mediated at least in part by salicylic acid and likely reflects salicylic acid inhibition of the synthesis and action of the wound signal jasmonic acid. We propose that the dual functions of salicylic acid contribute to a signal poise which constrains constitutive expression of disease and insect resistance mechanisms, and reciprocally switches their selective activation.

1999 Insect–plant interactions and induced plant defence. Wiley, Chichester (Novartis Foundation Symposium 223) p 166–174

Plants have two different inducible resistance mechanisms conferring systemic protection against pathogens and insect pests, respectively. Induction of systemic acquired resistance (SAR) to microbial and viral pathogens requires a necrotic

---

¶Author for correspondence

166

response on the infected leaf, leading to the transmission of a signal to distant leaves that triggers the systemic activation of an array of defences including deployment of pathogenesis-related (PR) proteins with direct antimicrobial activities (Ryals et al 1994). Induction of SAR is mediated by the phenylpropanoid-derived metabolite salicylic acid (SA) (Gaffney et al 1993, Delaney et al 1994). In contrast, insect damage activates a signal pathway involving systemin, a mobile 18-amino acid peptide, which stimulates the synthesis of the fatty acid-derived signal molecule jasmonic acid (JA), leading to the induction of proteinaceous antifeedants including various proteinase inhibitors and, in tobacco, synthesis of the alkaloid nicotine, a compound toxic to many insects (Baldwin et al 1994, Constabel et al 1995, Koiwa et al 1997, McConn et al 1997).

Induction of SAR can give immunity against a wide range of pathogens under field conditions (Ryals et al 1994), and likewise wound-induced pest resistance leads to increased plant performance (Agrawal 1998, Baldwin 1998). The effectiveness of these induced responses raises the question of why plants do not constitutively express SAR and systemic insect resistance mechanisms to reduce the susceptibility of naïve plants to attack by normally virulent pathogens and aggressive insect herbivores. To address this question we have examined grazing-induced insect resistance in transgenic tobacco plants with altered capacities for phenylpropanoid biosynthesis and SAR.

Phenylalanine ammonia-lyase (PAL) catalyses the first committed step in phenylpropanoid biosynthesis. Transgenic tobacco plants containing the bean *pal2* gene exhibit epigenetic co-suppression resulting in substantially reduced PAL activity and phenylpropanoid product accumulation. These PAL-suppressed plants are more susceptible to virulent fungal pathogens (Maher et al 1994) and fail to express SAR following infection by tobacco mosaic virus (TMV) (Pallas et al 1996). The latter may in part reflect the reduced levels of SA in these lines (Pallas et al 1996), because SAR is also compromised in transgenic tobacco expressing the bacterial *nahG* SA hydroxylase gene, which removes SA by converting it to catechol (Gaffney et al 1993, Delaney et al 1994).

In some *pal2* transgenic lines, co-suppression is lost following selfing and the resultant progeny exhibit several fold higher PAL activity and correspondingly elevated levels of chlorogenic acid (CGA, caffeoylquinic acid), the major phenylpropanoid product in tobacco leaves (Howles et al 1996). These PAL over-expressor plants show enhanced disease resistance. Thus, naïve plants are less susceptible to *Cercospora nicotianae* (S. V. Wesley & R. A. Dixon, unpublished results) and localized inoculation with TMV results in stronger induction of SAR, manifest by smaller lesions in response to subsequent viral challenge of distant leaves compared to the lesions in equivalent wild-type and PAL-suppressed plants (Felton et al 1999).

Primarily on the basis of studies with artificial diets, some phenolic compounds have been proposed to act directly as antifeedants (Elliger et al 1981). However, there is no significant difference in growth and survival of the insect pests *Heliothis viriscens* or *Manduca sexta* larvae following challenge of leaves of PAL over- or under-expressing transgenic tobacco plants that had not been induced by prior localized grazing (Bi et al 1997). In contrast, manipulation of PAL activity in transgenic tobacco has a marked and unexpected effect on the grazing-induced component of insect resistance (Felton et al 1999). Induced systemic resistance to *H. virescens* larvae can be conveniently measured by larval weight and mortality compared to respective control plants that have not been pre-induced by localized grazing on a lower leaf. By these assays, PAL-suppressed plants develop a more effective grazing-induced insect resistance response than wild-type plants. In contrast, systemic induced resistance to *H. virescens* is severely compromised in PAL over-expressor plants, with no grazing-inducible effect on larval weight and a substantially decreased incremental impact on mortality (Felton et al 1999). Thus, transgenic modulation of PAL expression has exactly opposite effects on the ability of tobacco to develop SAR and grazing-induced insect resistance.

Leaves of naïve PAL over-expressor plants have about twofold higher SA levels than equivalent leaves from healthy wild-type plants and the accumulation of SA in grazing-induced plants is substantially greater in the over-expressors. Moreover, PAL-suppressed plants show only a weak grazing-induced accumulation of SA from low basal levels. To test whether SA contributes to the reciprocal relationship between SAR and grazing-induced insect resistance, we repeated the insect feeding experiments with transgenic tobacco harbouring the bacterial *nahG* gene, and with its corresponding untransformed control line. *nahG* plants, which have drastically reduced levels of SA, fail to establish SAR in response to viral, fungal, or bacterial pathogens (Gaffney et al 1993, Ryals et al 1994). When naïve *nahG* plants are challenged with *H. virescens,* there are no significant differences in either larval weight or mortality compared to equivalent wild-type plants that have likewise not been pre-induced by localized grazing on a lower leaf. In contrast, systemic leaves of grazing-induced *nahG* tobacco are markedly more resistant to *H. virescens* than equivalent leaves in the corresponding wild-type control line. Thus SA contributes to the inverse relationship between the capacities for grazing-induced insect resistance and pathogen-induced SAR.

Transgenic suppression of *wipk*, encoding a wound-response MAP kinase, or *rgp1*, encoding a Rab-type GTP-binding protein, results in the wound-inducible accumulation of SA and SA-inducible transcripts encoding acidic PR proteins (Sano et al 1994, Seo et al 1995). While these data suggest mechanistic parallels, it is not clear whether in wild-type plants *wipk* or *rgp1* function as switches between the two pathways. However, exogenous SA blocks both JA biosynthesis and action in wound signalling (Doherty et al 1988, Peña-Cortés et al 1993, Doares et al 1995) and these pharmacological observations suggest a possible mechanism for the physiological cross-talk revealed by the transgenic perturbation of phenylpropanoid metabolism. Consistent with this idea, JA levels are inversely proportional to SA levels in both PAL transgenic and *nahG* plants (Felton et al 1999). This inverse relationship holds in both healthy, unwounded leaves and following wound induction. Moreover, genes encoding polyphenol oxidase and hydroxymethylglutaryl-CoA reductase, both of which are wound-inducible via the JA pathway (Baldwin et al 1994, Koiwa et al 1997, K. Korth, unpublished results), are most strongly expressed in systemic leaves of grazing-induced PAL-suppressed plants, whereas there is strong constitutive and insect-induced expression of the bean *pal2* transgene in the over-expressing lines, but not in the suppressed lines. Finally, nicotine levels, which are controlled in part via the JA pathway (Baldwin et al 1994), although systemically induced by insect grazing on PAL-suppressed plants, are not induced in the PAL over-expressors (Felton et al 1999).

The dual activity of SA as an inducer of SAR and suppressor of grazing-induced insect resistance means that impaired insect resistance is a constraint against the constitutive expression of SA-inducible disease resistance mechanisms. This negative cross-talk may play a role in prioritizing or channelling a plant's response to different biological stresses. Moreover, recent data indicate that exogenous JA inhibits SA induction of acidic PR proteins (Niki et al 1998), suggesting that pathway cross-talk may also operate in the opposite direction. Hence, the dual functions of SA may be part of a dynamic regulatory poise established in conjunction with JA. The enhanced grazing-induced insect resistance and elevated levels of JA observed in *nahG* plants imply that SA levels in uninfected wild-type plants are sufficient to give partial suppression of the grazing-induced pathway. Further accumulation of SA in response to attempted infection would decisively shift the poise toward induction of SAR and suppression of systemic insect resistance, whereas elevation of JA in response to insect grazing would be expected to favour induction of insect resistance at the expense of SAR.

Localized pathogen attack is an indicator that distant leaves are at high risk of subsequent infection but is not a predictor of insect attack. Likewise, localized insect grazing does not indicate a systemic risk of infection. Moreover, activation of disease resistance mechanisms involves massive, co-ordinated realignments of metabolic activity and gene expression distinct from those underlying induced insect resistance (Somssich & Hahlbrock 1998). Hence, in addition to imposing an additional energy burden with little selective advantage, co-induction of SAR and insect resistance mechanisms might be inherently inefficient and the SA/JA regulatory poise would help establish reciprocally selective activation of the two effector pathways. A corollary of SA-mediated negative cross-talk between SAR and systemic induced insect resistance is that agrochemical modulation or transgenic manipulation of signal pathway components upstream of the SA/JA nexus might have unwanted side effects in the field. However, such cross-talk may not be universal and indeed JA is a component of the resistance response of *Arabidopsis* roots to *Pythium* root rot (Vijayan et al 1998) and a SA-independent pathway for systemic disease resistance induced by root-colonizing *Pseudomonads* (Pieterse et al 1998).

Acknowledgements

This work was supported by the Samuel Roberts Noble Foundation and the USDA NRICGP.

References

Agrawal AA 1998 Induced responses to herbivory and increased plant performance. Science 279:1201–1202

Baldwin IT 1998 Jasmonate-induced responses are costly but benefit plants under attack in native populations. Proc Natl Acad Sci USA 95:8113–8118

Baldwin IT, Schmelz EA, Ohnmeiss TE 1994 Wound-induced changes in root and shoot jasmonic acid pools correlate with induced nicotine synthesis in *Nicotiana sylvestris*. J Chem Ecol 20:2139–2157

Bi JL, Felton GW, Murphy JB, Howles PA, Dixon RA, Lamb CJ 1997 Do plant phenolics confer resistance to specialist and generalist insect herbivores? J Agric Food Chem 45:4500–4504

Constabel CP, Bergey DR, Ryan CA 1995 Systemin activates synthesis of wound-inducible tomato leaf polyphenol oxidase via the octadecanoid defense signaling pathway. Proc Natl Acad Sci USA 92:407–411

Delaney TP, Uknes S, Vernooij B et al 1994 A central role of salicylic acid in plant disease resistance. Science 266:1247–1250

Doares SH, Narváez-Vásquez J, Conconi A, Ryan CA 1995 Salicylic acid inhibits synthesis of proteinase inhibitors in tomato leaves induced by systemin and jasmonic acid. Plant Physiol 108:1741–1746

Doherty HM, Selevendran RR, Bowles DJ 1988 The wound response of tomato plants can be inhibited by aspirin and related hydroxybenzoic acids. Physiol Mol Plant Pathol 33:377–384

Elliger CA, Wong Y, Chan BG, Waiss AC Jr 1981 A growth inhibitor in tomato (*Lycopersicon*) to tomato fruitworm (*Heliothis zea*). J Chem Ecol 4:753–758

Felton GW, Bi JL, Korth KL et al 1999 Inverse relationship between systemic resistance of plants to microorganisms and to insect herbivory. Curr Biol, in press

Gaffney T, Friedrich L, Vernooij B et al 1993 Requirement of salicylic acid for the induction of systemic acquired resistance. Science 261:754–756

Howles PA, Sewalt VJH, Paiva NL et al 1996 Overexpression of L-phenylalanine ammonia-lyase in transgenic tobacco plants reveals control points for flux into phenylpropanoid biosynthesis. Plant Physiol 112:1617–1624

Koiwa H, Bressan RA, Hasegawa PM 1997 Regulation of protease inhibitors and plant defense. Trends Plant Sci 2:379–384

Maher EA, Bate NJ, Ni W, Elkind Y, Dixon RA, Lamb CJ 1994 Increased disease susceptibility of transgenic tobacco plants with suppressed levels of preformed phenylpropanoid products. Proc Natl Acad Sci USA 91:7802–7806

McConn M, Creelman RA, Bell E, Mullet JE, Browse J 1997 Jasmonate is essential for insect defense in *Arabidopsis*. Proc Natl Acad Sci USA 94:5473–5477

Niki T, Mitsuhara I, Seo S, Ohtsubo N, Ohashi Y 1998 Antagonistic effect of salicylic acid and jasmonic acid on the expression of pathogenesis-related (PR) protein genes in wounded mature tobacco leaves. Plant Cell Physiol 39:500–507

Pallas JA, Paiva NL, Lamb CJ, Dixon RA 1996 Tobacco plants epigenetically suppressed in phenylalanine ammonia-lyase expression do not develop systemic acquired resistance in response to infection by tobacco mosaic virus. Plant J 10:281–293

Peña-Cortés H, Albrecht T, Prat S, Weiler EW, Willmitzer L 1993 Aspirin prevents wound-induced gene expression in tomato leaves by blocking jasmonic acid biosynthesis. Planta 191:123–128

Pieterse CMJ, van Wees SCM, van Pelt JA et al 1998 A novel signaling pathway controlling induced systemic resistance in *Arabidopsis*. Plant Cell 10:1571–1580

Ryals J, Uknes S, Ward E 1994 Systemic acquired resistance. Plant Physiol 104:1109–1112

Sano H, Seo S, Orudgev E, Youssefian S, Ishizuka K, Ohashi Y 1994 Expression of the gene for a small GTP binding protein in transgenic tobacco elevates endogenous cytokinin levels, abnormally induces salicylic acid in response to wounding, and increases resistance to tobacco mosaic virus infection. Proc Natl Acad Sci USA 91:10556–10560

Seo S, Okamoto M, Seto H, Ishizuka K, Sano H, Ohashi Y 1995 Tobacco MAP kinase: a possible mediator in wound signal transduction pathways. Science 270:1988–1992

Somssich IE, Hahlbrock K 1998 Pathogen defence in plants—a paradigm of biological complexity. Trends Plant Sci 3:86–90

Vijayan P, Shockey J, Levesque CA, Cook RJ, Browse J 1998 A role for jasmonate in pathogen defense of *Arabidopsis*. Proc Natl Acad Sci USA 95:7209–7214

DISCUSSION

Hammerschmidt: At the beginning of your paper you described the systemic induction of resistance after the incompatible interaction. Don't you see the systemic response with necrosis induced by a compatible pathogen?

Lamb: What Ray Hammerschmidt is referring to is work showing that certain necrotizing pathogens will also induce SAR. We have never looked at that kind of system in *Arabidopsis*.

Hammerschmidt: In cucumber we get a much better systemic response with the virulent necrotic-lesion-causing pathogens as opposed to the avirulent hypersensitive response-inducing pathogens.

Lamb: I think that reflects the biology of the initial pathogen attack: a pathogen coming in that is necrotizing from an early stage causes strong necrosis locally, and will be an effective inducer of SAR, but we have never tested that in *Arabidopsis*.

Baldwin: Your *pal*-suppressed plants presumably have low CGA levels (which is a main substrate for the polyphenol oxidases [PPOs] that are activated by JA). Does the lack of induced response to caterpillar attack in *pal*-suppressed plants speak to the lack of importance of substrate supply in PPO-based defences?

Lamb: I think so. Gary Felton has found that in the naïve plants, the lack of chlorogenic acid does not affect the outcome of the herbivory (Felton et al 1999). The only effect we see with insects is when we look at the grazing-induced response. We have no data arguing for or against the PPO-mediated defences.

Schultz: An ecological note: in that system, CGA is a substrate for PPOs, and one of its major effects is to inhibit viruses that kill the caterpillar. The tritrophic consequences of this would be greater susceptibility to pathogens by the insects.

I was curious: *nahG* works by rapidly metabolizing salicylate. Does the outcome of your experiments with those plants necessarily imply an iterative process or merely waiting long enough for the salicylate that's produced to have been used up?

Lamb: By iterative processes, what I mean is that we see an oxidative burst at the primary site of inoculation, but we also see oxidative bursts at a distance, subsequently. The pharmacology suggests that the primary burst is driving the secondary microbursts. In principle, the secondary microbursts can drive tertiary microbursts: in fact, we have some evidence for that from the pattern of glutathione *S*-transferase expression, where you can actually see waves of induction. We therefore have the problem of what is triggering the secondary burst. A further issue is that SA seems to function both upstream and downstream of the oxidative burst, depending on how we do the experiment. That is, there are many data showing that to get a strong oxidative burst you need SA. Then, if you mimic the oxidative burst, you can induce SA. We are staring to think that there is a cyclical process that requires both peroxide and salicylate. The somewhat conflicting data in the literature are caused by different experimental designs.

Some of these lesion mimic mutants seem to operate upstream of cell death and others seem to operate downstream of SA and so on. The Ciba group have proposed that this is cyclical too. My view of the Ciba data is that you could explain them by two parallel pathways. Our data are harder to explain in that way.

Pickett: I have a technical question that has troubled me ever since I saw the *nahG* gene work from John Ryals' lab (Gaffney et al 1993). This gene is for an enzyme that gives an oxidative decarboxylation of salicylate. How do you accommodate the fact that this system is using up some oxidizing capacity?

Lamb: I don't think anyone has addressed that. It is not just a null mutation. There are consequences of *nahG* expression beyond just removing salicylate. The one that people have focused on is catechol production, but it has been carefully demonstrated that catechol does not account for compromised resistance.

Dietrich: By feeding ^{14}C-labelled salicylate to *nahG* plants and using an HPLC detection system, John Draper has shown that although the product of the *nahG* gene converts salicylate to catechol, catechol does not accumulate in the *nahG* plants (Mur et al 1997).

Lamb: One of the issues with the *nahG* story is that these experiments have been done almost exclusively with the cauliflower mosaic virus 35S promoter, which doesn't deliver strong expression everywhere. There are suggestions from some of our work where we see these periveinal responses (there's a lot of action going on in the cells surrounding the vein), that salicylate could be a mobile signal because the *nahG* isn't being delivered very effectively in some of the cells that count. John Draper has gone back and put *nahG* under the control of a phloem-specific promoter, and there it gives a much stronger block of SAR. I think he's done grafting experiments which show that salicylate could be a mobile signal.

Noel Keen has also observed that a very early event is rapid PAL induction in the petiole of the emitting leaf. His argument is that this might then be a salicylate-generating factory.

Pickett: Peroxide can move around as an organic peroxide; there is considerable potential for such compounds in these systems.

Lamb: Those are the kind of metabolites one might be interested in.

Schultz: This opens several doors to manipulating wound responses by insects. The one that's newest — and about which I know the least — is manipulating nitric oxide synthase (NOS). Having done all this work with all these inhibitors, are there natural products or processes that would be available for blocking that step?

Lamb: There are some nitrosylated phenols that accumulate in plants, and one might look at these for biological activity. Nitric oxide might react with superoxide, but there are some data from Massimo Delledonne, who did the original nitric oxide work, that would suggest this is not the way it is functioning in plants. It is rather a complicated argument. We've been racking our brains to think of a transgenic approach for confirming the pharmacological work with nitric oxide. Dan Klessig has shown that you can augment plants with purified NOS (Durner et al 1998). Another interesting new development is a recent paper by Gardner et al (1998) that shows that flavohaemoglobin is a nitric oxide dioxygenase. In the discussion of this paper, the authors suggest that this may explain why non-leguminous plants produce haemoglobin. Secondly, they noted that there's a mutant in *Erwinia* in which the flavohaemoprotein is knocked out, and this mutant is less virulent on plants.

Scheel: I was surprised about data from tobacco transgenics, where you had this differential response with the antagonistic effect of salicylate and jasmonate. In *Arabidopsis* infected with *Fusarium* or *Pseudomonas* you see both increased salicylate and jasmonate: are these separated?

Lamb: This is something that I worry about. It's clear that jasmonate in certain circumstances, probably in a more limited set of interactions, functions in disease resistance. Perhaps it functions in response to necrotrophic pathogens, which may be reminiscent of a wound response.

References

Durner J, Wendehenne D, Klessig DF 1998 Defense gene induction in tobacco by nitric oxide, cyclic GMP, and cyclic ADP-ribose. Proc Natl Acad Sci USA 95:10328–10333

Felton G, Bi JL, Korth KL et al 1999 Inverse relationship between systemic resistance of plants to microorganisms and to insect herbivory. Curr Biol, in press

Gaffney T, Friedrich L, Vernooij B et al 1993 Requirement of salicylic acid for the induction of systemic acquired resistance. Science 261:754–756

Gardner PR, Gardner AM, Martin LA, Salzman AL 1998 Nitric oxide dioxygenase: an enzymic function for flavohemoglobin. Proc Natl Acad Sci USA 95:10378–10383

Mur LAJ, Bi Y-M, Darby RM, Firek S, Draper J 1997 Compromising early salicylic acid accumulation delays the hypersensitive response and increases viral dispersal during lesion establishment in TMV-infected tobacco. Plant J 12:1113–1126

The role of phytoalexins in plant protection

R. Hammerschmidt and E. K. Dann*

*Department of Botany and Plant Pathology, Michigan State University, East Lansing, MI 48824-1312, USA and *Department of Crop Sciences, University of Sydney, Sydney, NSW 2006, Australia*

Abstract. Plants resist pathogens with combinations of constitutive and induced defences. Of the induced defences, phytoalexin production has received much attention since the phytoalexin concept was introduced over 50 years ago. However, the specific role of phytoalexins in disease resistance is not clear for the majority of host–parasite systems. Much of the research on phytoalexins has relied on the identification of induced antifungal compounds and correlating their presence with resistance. Although an important first step, more definitive studies are needed. Approaches that use *in situ* localization and quantification have provided good evidence that phytoalexins can accumulate at the right time, concentration, and location to be effective in resistance. Studies on phytoalexin tolerance in pathogenic fungi have also shown a relationship between virulence and the ability of fungi to detoxify phytoalexins. Use of mutants deficient in phytoalexin synthesis and elucidating biosynthetic pathways provide other approaches to evaluating the role of phytoalexins. These approaches will be illustrated using the *Arabidopsis*–camalexin system and other selected examples. Studies on natural variation in phytoalexin production and the accumulation of phytoalexins under field conditions will be presented as another means of evaluating the role of these compounds in resistance to pathogens and other pests.

1999 Insect–plant interactions and induced plant defence. Wiley, Chichester (Novartis Foundation Symposium 223) p 175–190

Phytoalexin production is one of the most studied of several defence responses expressed after infection. Although the phytoalexin concept was first proposed nearly 60 years ago (Muller & Borger 1940), the relative contribution of phytoalexins to disease resistance is still not clearly established. Because of the lack of a clearly defined role for phytoalexins, the current definition of phytoalexins as antimicrobial compounds produced after infection (Kuc 1995) has been generally accepted in lieu of a definition that implicates these compounds in defence.

Phytoalexins have been characterized from 31 flowering plant families and represent a diverse array of secondary metabolites (Grayer & Harborne 1994). Representative examples are shown in Fig. 1. Although most plants within a family produce structurally related phytoalexins (e.g. sesquiterpenes in Solanaceae), this pattern does not always follow as illustrated by the different chemical classes that are found in the graminaceous monocots (Grayer & Harborne 1994).

Most early work on phytoalexins involved characterizing induced antifungal compounds and showing that the phytoalexins accumulated more quickly and to antimicrobial concentration in resistant as compared to susceptible interactions

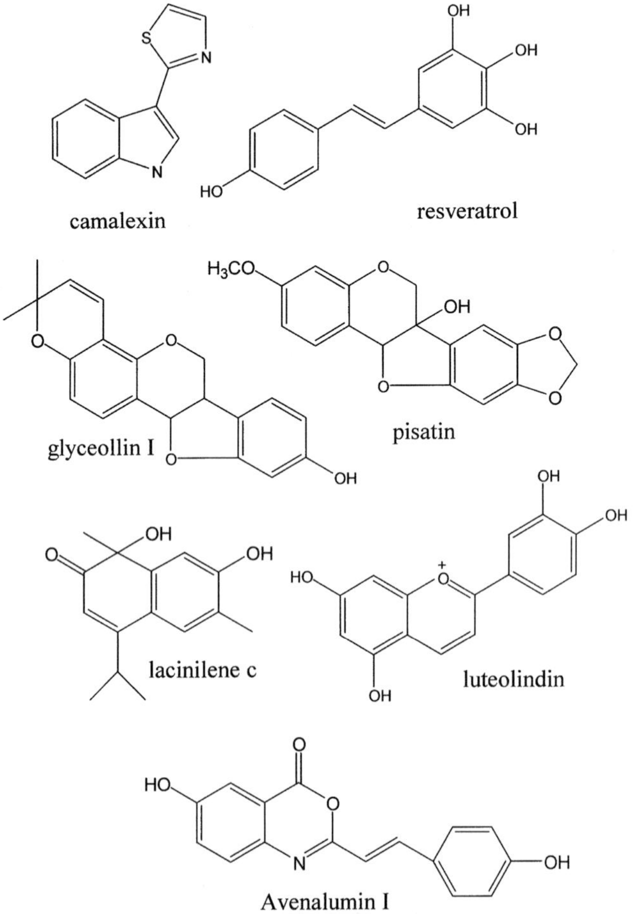

FIG. 1.   Representative phytoalexins discussed in text. Camalexin from *Arabidopsis*, *Camalina* and *Arabis* (Tsuji et al 1992, Browne et al 1991, Zook et al 1998); resveratrol from grapevine and peanut (Hahn et al 1993, Thomzik et al 1997); glyceollin I from soybean (Hahn et al 1985; E. K. Dann & R. Hammerschmidt, unpublished work); pisatin from garden pea (VanEtten et al 1989); lacinilene c from cotton (Pierce et al 1996); luteolindin from sorghum (Snyder et al 1991); avenalumin I from oats (Mayama & Tani 1982).

(Cruickshank 1963). These types of data, although useful for preliminary assessment, do not provide the type of conclusive evidence needed to demonstrate the contribution of phytoalexins to resistance. In this paper, we will discuss some of the newer evidence and approaches for demonstrating the role of phytoalexins in defence as well as some of our recent work on the role of phytoalexins in resistance of *Arabidopsis* and soybean to disease.

Localization studies

To be effective, phytoalexins must accumulate to inhibitory concentrations at the site of pathogen development (Nicholson & Hammerschmidt 1992). One approach to satisfy these conditions is to localize the phytoalexin with respect to the pathogen via cytological means, and some successes with this have been reported.

Infection of sorghum seedlings with *Colletotrichum graminicola*, a maize pathogen, results in accumulation of red-to-orange coloured 3-deoxyanthocyanidin phytoalexins luteolindin (Fig. 1), apigeninidin and a caffeic acid ester of arabinosyl-5-O-apigeninidin (Nicholson et al 1987, Snyder et al 1991). Snyder & Nicholson (1990) showed cytologically that these compounds develop in inclusion bodies that form in the cytoplasm of the plant cell being infected. The first colourless inclusion could be seen in epidermal cells associated with fungal appressoria at approximately the time the fungal appressorium was maturing. These inclusions migrate towards the infection peg and, as the migration occurs, the phytoalexins accumulate in the inclusions. Eventually the inclusions coalesce and the phytoalexins are released into the cytoplasm. Using microspectrophotometry, Snyder et al (1991) quantified the amount of phytoalexin in the infected host cells. The concentration of all three phytoalexins (based on luteolindin equivalents) in an inclusion was reported to be 150 μM. HPLC analysis showed that each infected plant cell accumulated 0.48–1.20 ng luteolindin and 0.24–0.91 ng apigeninidin per cell. These data indicated that the amount of phytoalexin accumulating in individual cells exceeded the amount needed for *in vitro* toxicity (*c.* 9 μM) and provided very good evidence for a role of these 3-deoxyanthocyanidin phytoalexins in host resistance.

Timing and cellular localization of studies have also provided evidence that supports a role for phytoalexins in the resistance of cotton to *Xanthomonas campestris* (e.g. lacinilene c, Fig. 1; Pierce et al 1996), oats to *Puccinia coronata* (e.g. avenalumin I, Fig. 1; Mayama & Tani 1982) and soybean to *Phytophthora megasperma* (e.g. glyceollin I, Fig. 1; Hahn et al 1985).

Critical to all of the localization/timing studies is a good correlation between the amount of phytoalexin that likely is in contact with the pathogen and *in vitro* toxicity. Unfortunately, no studies have been able to directly show that the phytoalexin responsible for stopping the pathogen or that the phytoalexin is actually toxic to the pathogen in the tissue. Use of pathogen mutants that are resistant to the phytoalexin or, as described below, selection of natural variants of the pathogen that are phytoalexin tolerant may provide a means to demonstrate the relationship between *in vitro* toxicity, *in planta* levels of phytoalexins, and the contribution of the phytoalexin to defence.

Pathogen tolerance to phytoalexin and virulence

VanEtten and co-workers have shown that part of the virulence of *Nectria hematococca* to peas is based on the ability of the pathogen to detoxify pisatin (Fig. 1) (early work reviewed in VanEtten et al 1989). These studies originated from the observation that natural isolates of *N. hematococca* that were pathogenic on pea could also detoxify pisatin (Fig. 1) *in vitro*. Genetic analysis indicated that pathogenicity segregated with pisatin detoxification, and this led to the discovery of the enzyme pisatin demethylase and a set of genes encoding this enzyme. Disruption of the *pda* gene (Wasmann & VanEtten 1996) or transformation of a *pda⁻* strain with pisatin demethylase (Ciufetti & VanEtten 1996) only resulted in a small decrease or increase, respectively, in virulence. This result was explained by the location of the *pda* gene on a dispensable chromosome that has other pathogenicity genes (VanEtten et al 1994). *Nectria* isolates that were not pea pathogens and could not detoxify pisatin did not have this chromosome. Even though the original genetic data suggested that pisatin detoxification was a major pathogenicity factor, subsequent work has demonstrated that other factors are also important in the defence response of pea plants to *N. hematococca*. When considered along with all other reported defences (Hammerschmidt & Schultz 1996), it is not a surprising that pisatin only partially contributes to the resistance. However, the research by VanEtten and co-workers has led to a better understanding of the regulation of virulence in this necrotroph and the identification of other genes that are required for pathogenicity.

Molecular manipulation of the phytoalexin response

Another approach to studying the role of phytoalexins is to modify the plant to produce higher levels of or novel phytoalexins. Transformation of tobacco and tomato with the gene for resveratrol synthase resulted in accumulation of the phytoalexin resveratrol (Fig. 1) and enhanced disease resistance in several plants. Transformed tobacco was somewhat more resistant to infection by *Botrytis cinerea* (Hain et al 1993). More encouraging results were found with resveratrol synthase transformed tomato challenged with *Phytophthora infestans* (Thomzik et al 1997).

These results show that enhancing levels of phytoalexins can enhance disease resistance and, therefore, may be of value in crop protection.

Engineering plants to produce secondary metabolites such as phytoalexins may not always be as straightforward as resveratrol. Many of the phytoalexins (e.g. the isoflavonoids and pterocarpans from legumes) have rather complex biosynthesis (Kuc 1995). However, even in simple cases, the transformed plants may produce compounds in addition to the desired end product. Tobacco and other solanaceous plants produce sesquiterpenoid phytoalexins which are produced following conversion of farnesyl pyrophosphate into a sesquiterpene hydrocarbon which is further modified (Kuc 1995). Transformation of tobacco with trichodiene synthase, a sesquiterpene cyclase from the fungus *Fusarium sporotrichiodes*, resulted in transformed plants producing both the active cyclase enzyme and trichodiene, the product of this enzyme. However, 5-hydroxy-trichodiene, a compound found in neither tobacco nor the fungus, was found in the transformed tobacco cells (Zook et al 1996a,b). Thus, when examining the production of plants engineered to produce novel phytoalexins, the plant tissues should be analysed for not only the desired end-product, but also new compounds that may be derived from them.

Use of phytoalexin mutants to assess the role of phytoalexins

A logical approach to evaluating the role of phytoalexins in defence is to use mutants that do not produce these compounds. Potential difficulty arises with mutants because most of the plant species that produce phytoalexins are capable of synthesizing multiple phytoalexins (Kuc 1995, Grayer & Harborne 1994). The discovery that *Arabidopsis* appears to produce only camalexin (Tsuji et al 1992, Fig. 1) suggested that this plant may provide the system that would allow a mutational analysis of phytoalexins in plant defence.

Glazebrook & Ausubel (1994) reported the isolation of several *Arabidopsis* mutants that were phytoalexin deficient (*pad* mutants). The first three mutants isolated, *pad1–1*, *pad2–1* and *pad3–1,* were found to produce 30%, 10% and non-detectable amounts of camalexin, respectively, as compared to the wild-type in response to inoculation with the *Arabidopsis* pathogen *Pseudomonas syringae* pv. *maculicola.* None of the mutants were more susceptible to infection by incompatible *P. syringae* pathovars, although *pad1–1* and *pad2–1* did show greater susceptibility to *P. syringae* pv. *maculicola.* Two other mutants, *pad4* and *pad5* were isolated in a screen with *P. syringae* pv. *maculicola* and the maize pathogen *Cochliobolus carbonum*, respectively. Even though *pad5* was found in response to *C. carbonum* (a very good inducer of camalexin, Zook & Hammerschmidt 1997), *pad5* plants were no less resistant to *C. carbonum* than the wild-type. Comparisons of camalexin accumulation in the mutants by *Pseudomonas* or *Cochliobolus* have produced interesting results. For example, *pad4* produces wild-type levels of camalexin in response to *Cochliobolus* while the same plant only produces about 10% of wild-type in response to *Pseudomonas syringae*. A summary of the accumulation of camalexin by various *pad* mutants in response to fungal or bacterial infection can be found in Table 1.

Several *pad* mutants were tested for their effect on infection by incompatible isolates of *Peronospora parasitica* (Glazebrook et al 1997). *Pad4–1* mutants expressed enhanced susceptibility to four of the six pathogen isolates and other mutant–*Peronospora* combinations showed enhanced susceptibility (Table 1). *Pad5–1* and *pad2–1*, which produce very low levels of camalexin in response to *C. carbonum*, showed no or only a slight increase in susceptibility against two *Peronospora* isolates. Unfortunately, no measurements of camalexin content in the *Peronospora*-infected plants was reported. Thus, it is difficult to assess the role of camalexin in this specific interaction.

Elicitation of much more camalexin in *pad4* by *C. carbonum* than *Pseudomonas* suggested that *pad4* might be a regulatory rather than biosynthetic gene. The *pad4* plants do not express the PR1 gene nor accumulate salicylic acid to the same level as the wild-type after *P. syringae* pv. *maculicola* infection (Zhou et al 1998). Pretreatment of *pad4* plants with salicylic acid prior to inoculation did enhance the ability of the plants to accumulate camalexin to near wild-type levels. Similar to the results obtained with *C. carbonum*, inoculation of *pad4* with avirulent strains of *P. syringae* resulted in near wild-type levels of camalexin, thus further supporting a regulatory role for this gene (Zhou et al 1998).

It is possible that the other *pad* genes may also be regulatory in nature rather than biosynthetic. A labelling study with [^{14}C]anthranilic acid, a precursor of

TABLE 1   Characteristics of *Arabidopsis pad* mutants

| Mutant | Camalexin induction (% of wild type) | | |
	P.s. maculicola	*C. carbonum*	*P. parasitica sporulation*[a]
pad1	30	30	None (all 6)
pad2	10	10	Low/med (1); rare (1); none (4)
pad3	<1	—	Low (1); rare (1); none (4)
pad4	10	100	Heavy (3); med/heavy (1); none (2)
pad5	10	10	None (6)

[a]All six isolates of *P. parasitica* are incompatible on wild-type (Columbia) *Arabidopsis*. Numbers after sporulation rating refer to number of isolates out of the six that fell into each rating.
Data summarized from Glazebrook & Ausubel (1994) and Glazebrook et al (1997).

camalexin, did not reveal the accumulation of any putative precursors in several *pad* mutants (I. A. Kagan & R. Hammerschmidt, unpublished results). Interestingly, the labelling study also did not reveal any major difference in the amount of label in camalexin produced by *pad1* as compared to wild-type. These results further suggest that the *pad* genes may be regulatory rather than biosynthetic. Thus, if these genes do regulate resistance mechanisms, expression of other defences in the mutants needs to be evaluated. The overall conclusions from these studies, coupled with relatively low accumulation levels, indicate that camalexin is not a critical factor in defence against bacteria. This is supported by the analysis of Mansfield (1999) who pointed out the low antibacterial activity of camalexin (Tsuji et al 1992, Rogers et al 1996). Similarly, the lack of disease development by *C. carbonum* on *pad* mutants suggests that non-host resistance does not rely exclusively on camalexin accumulation.

Natural variation in phytoalexin production

Another approach to evaluating the role of phytoalexins could be to take advantage of natural variability in phytoalexin production in a plant species. For example, greater disease resistance of wild as compared to domesticated *Phaseolus coccineus* was correlated with higher total amounts and a greater diversity of isoflavonoids produced after elicitation with $CuCl_2$ (Lindig-Cisneros et al 1997) The same resistance trend was found with *Phaseolus lanutus*, but this could not be explained by phytoalexin diversity or accumulation (the wild *P. lanutus* were, however, more cyanogenic). Evaluation of rice germplasm also showed a correlation between blast resistance and phytoalexin accumulation (Dillon et al 1997). It is, however, difficult to assess whether the phytoalexin accumulation in these cases is directly related to the defence or whether it is just a response.

To a limited extent, we have shown that resistance of soybean to the necrotrophic, broad host-range pathogen *Sclerotinia sclerotiorum*, is correlated with the relative levels of induction of glyceollin by an abiotic elicitor (Table 2). Although these results do not conclusively implicate glyceollin in defence, rapid expression of a defence such as glyceollin would be necessary in the resistance to a generalist necrotrophic pathogen like *Sclerotinia*. We have examined a number of *Arabidopsis* ecotypes for their ability to produce camalexin (I. A. Kagan & R. Hammerschmidt, unpublished results). Some variation has been detected, and this may be useful in future analysis of the role of camalexin in defence.

Camalexin biosynthetic studies

A plant that is blocked at a step in phytoalexin synthesis without disruption of other defences or processes would be an ideal system for evaluating the role of

TABLE 2   Effect of plant activators on induction of glyceollin

| Pre-treatment | Glyceollin accumulation after inoculation[a] | |
	12 hours	24 hours
Water	0.152	1.223
INA	0.174	2.302
BTH	0.170	3.044

[a]Leaves were inoculated with *C. carbonum*. Glyceollin content expressed as $\mu g/g$ fresh weight.

phytoalexins. Since mutants have not led to the discovery of such a biosynthetic mutant, an alternative approach has been taken. A biosynthetic pathway has been proposed for camalexin starting with anthranilic acid with a branch point prior to tryptophan (Fig. 2). Labelling and mutant studies have shown that anthranilic acid, but not the anthranilic acid metabolite tryptophan, is a precursor of camalexin (Tsuji et al 1993). We have also shown that the thiazole ring of camalexin is derived from the cyclization of cysteine (Zook & Hammerschmidt 1997). Studies with cell suspension cultures have recently shown that free indole may be a precursor (Zook 1998). One proposed biosynthetic pathway suggests that indole-3-carboxaldehyde is a precursor that condenses with cysteine to form the immediate precursors to camalexin (Browne et al 1991). We have demonstrated that indole-3-carboxaldehyde is present in *Arabidopsis*, and that [^{14}C]anthranilic acid is incorporated into this aldehyde. The results are supported by the activation of tryptophan biosynthetic genes during the elicitation of camalexin (Zhao & Last 1996). The goal of this work is to elucidate the pathway, and then use the precursors to search for the enzymes (and eventually the genes) that control the camalexin biosynthetic steps. Ultimately, camalexin-minus *Arabidopsis* and (and phytoalexin-minus plants of other species) would further our understanding of the role of phytoalexins.

Induced resistance and phytoalexins

The original concept of the phytoalexin was derived from experiments on the local induction of resistance of potato tuber tissue by incompatible races of *Phytophthora infestans* to compatible *P. infestans* and a tuber rotting *Fusarium* species (Muller & Borger 1940). However, few studies since then have correlated the production of

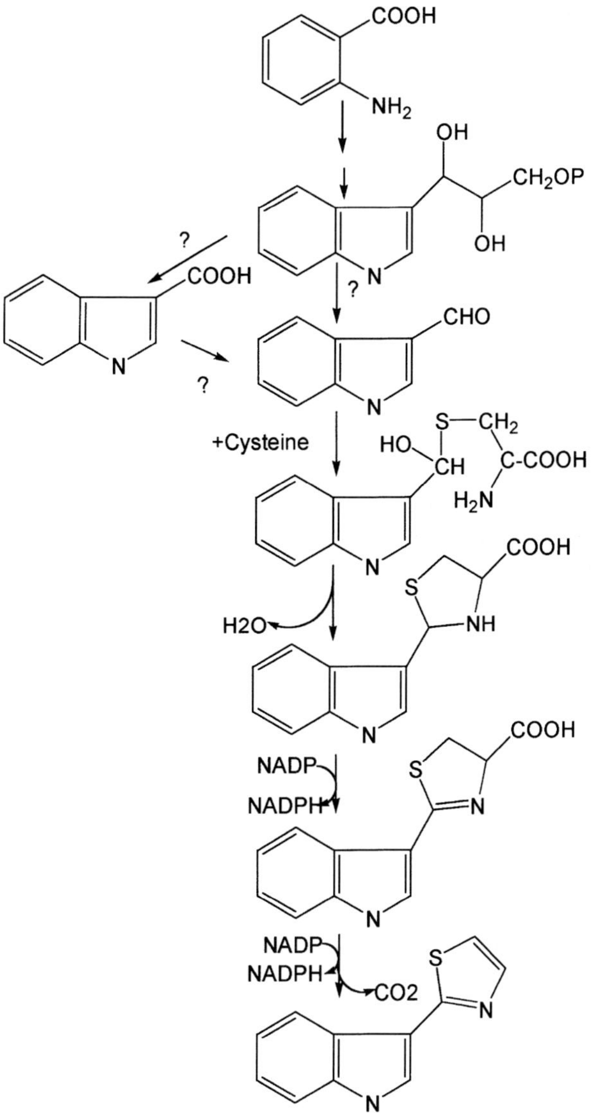

FIG. 2. Hypothetical biosynthetic pathway for camalexin (modified from Zook & Hammerschmidt 1997).

phytoalexins with local induced resistance or with the ability of systemically induced tissues to produce phytoalexins after challenge. We have recently examined the role of phytoalexins in the induced resistance response of soybean to the white mould pathogen *Sclerotinia sclerotiorum*.

Treatment of field-grown soybean plants with the resistance-activating compounds INA (2,6-dichloroisonicotinic acid) or BTH (benzo(1,2,3)thiadiazole-7-carbothioic acid S-methyl ester) reduces the amount of white mould by up to or over 50% (Dann et al 1998). Analysis of foliage treated with BTH in the field or greenhouse showed, as expected, that the plant activator did not elicit the accumulation of glyceollin (Table 2). However, inoculation of the foliage with *Cochliobolus carbonum*, a pathogen of maize, showed that the rate and amount of glyceollin accumulation was greater in BTH- or INA-treated plants than water controls (Table 3). Thus, it is possible that a more rapid accumulation of glyceollin in BTH/INA pre-treated plants in response to *S. sclerotiorum* may account for the enhanced resistance seen in field and greenhouse trials. This possibility is now being investigated.

We have also found another type of induced resistance that provides some evidence for the ability of glyceollin to reduce white mould of soybeans. Observations made by soybean growers suggested that post-emergent treatment of the crop with certain diphenylether herbicides such as lactofen reduced the amount of white mould. In replicated trials over several years, we were able to show that under conditions of high disease pressure, application of lactofen reduced the amount of white mould if applied at specific growth stages of the plant. Analysis of lactofen-treated leaves revealed that they contained significantly higher levels of glyceollin than controls in both greenhouse and field studies in 1996 (Table 3) and 1997 (data not shown). Treated leaves were also more resistant to *Sclerotinia* infection than the controls (the amount of lacotofen used was not toxic to the pathogen). Analysis of untreated, systemic leaves from field-treated plants several weeks later revealed a reduction of glyceollin content in the leaves and a loss of resistance of the leaves to white

TABLE 3   White mould resistance and accumulation of glyceollin in field-grown soybean leaves treated with lactofen or BTH in 1996

Cultivar	Treatment	Glyceollin[a]	Lesion size[b]
NKS-1990	Water	0	10.9
	BTH	0	10.7
	Lactofen	7.4*	6.5*
Williams 82	Water	0	10.2
	BTH	0.3	8.5*
	Lactofen	38.2*	4.6*

[a]Glyceollin content: μg/g fresh weight at three days after treatment
[b]Lesion size (mm) caused by *S. sclerotiorum* on detached soybean leaves of the various treatments.
*Significantly different from control.

mould. These observations suggest that glyceollin, if induced in tissue to fungitoxic levels, may add to the resistance of the plant. However, these data do not exclude the possible role of other defences that were induced by lactofen.

The herbicidal mode of action of lactofen and related diphenyl ether herbicides may provide a clue to its glyceollin-eliciting activities. Lactofen and other *p*-nitro-diphenylethers inhibit the enzyme protophorphyrinogen IX oxidase. This results in accumulation of protoporphyrinogen IX which can be oxidized to the protoporphyrin. In the light, protoporphyrin generates singlet oxygen (Scalla & Matringe 1994), which may contribute to glyceollin elicitation (Fig. 3).

Conclusions

When viewed collectively, there is good evidence that phytoalexins play some role in defence against pathogens. However, most of the evidence is correlative and does not truly determine cause and effect or the interaction of phytoalexins with other putative defences. The use of mutants and the ability to engineer plants with modified secondary metabolism provides new opportunities to determine more precisely the relative contribution of phytoalexins to disease resistance and thus provides a logical framework to exploit these natural products in plant protection.

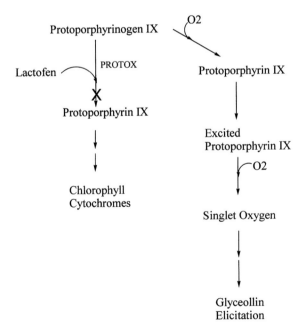

FIG. 3.   Possible mode of action of the *p*-nitro-diphenyl ether herbicide lactofen in glyceollin elicitation. Modified from Scalla & Matringe (1994) to illustrate glyceollin elicitation. Lactofen, a diphenylether herbicide, inhibits the enzyme protoporphyrinogen oxidase (PROTOX).

Acknowledgements

The support of the National Science Foundation (IBN-9220912), the Michigan Soybean Promotion Committee, and the Michigan Agricultural Experiment Station is gratefully acknowledged.

References

Browne LM, Conn KL, Ayer WA, Tewar JP 1991 The camalexins: new phytoalexins produced in the leaves of *Camelina sativa* (Cruciferae). Tetrahedron 47:3909–3914

Ciuffetti L, VanEtten H 1996 Virulence of a pisatin demethylase-deficient *Nectria haematococca* MPVI isolate is increased by transformation with a pisatin demethylase gene. Mol Plant-Microbe Interact 9:787–792

Cruickshank IAM 1963 Phytoalexins. Annu Rev Phytopathol 1:351–374

Dann EK, Diers B, Byrum J, Hammerschmidt R 1998 Effect of treating soybean with 2,6-dichloroisonicotinic acid (INA) and benzothiadiazole (BTH) on seed yields and the level of disease caused by *Sclerotinia sclerotiorum* in field and greenhouse studies. Euro J Plant Path 104:271–278

Dillon VM, Overton J, Grayer RJ, Harborne JB 1997 Differences in phytoalexin response among rice cultivars of different resistance to blast. Phytochemistry 44:599–603

Glazebrook J, Ausubel FM 1994 Isolation of phytoalexin-deficient mutants of *Arabidopsis thaliana* and characterization of their interactions with bacterial pathogens. Proc Natl Acad Sci USA 91:8955–8959

Glazebrook J, Zook M, Mert F et al 1997 Phytoalexin-deficient mutants of *Arabidopsis* reveal that *PAD4* encodes a regulatory factor and that four *PAD* genes contribute to downy mildew resistance. Genetics 146:381–392

Grayer RJ, Harborne JB 1994 A survey of antifungal compounds from plants, 1982–1993. Phytochemistry 37:19–42

Hahn MG, Bonhoff A, Griesbach H 1985 Quantitative localization of the phytoalexin glyceollin I in relation to fungal hyphae in soybean root infected with *Phytophthora infestans* f. sp. glycinea. Plant Physiol 77:591–601

Hain R, Reif H, Krause E et al 1993 Disease resistance results from foreign phytoalexin expression in a novel plant. Nature 361:153–156

Hammerschmidt R, Schultz J 1996 Multiple defenses and signals in plant defense against pathogens and herbivores. In: Romeo JT, Saunders JA, Barbosa P (eds) Phytochemical diversity and redundancy in ecological interactions. Plenum, New York (Rec Adv Phytochem 30) p 121–154

Kuc J 1995 Phytoalexins, stress metabolism and disease resistance in plants. Annu Rev Phytopathol 33:275–297

Lindig-Cisneros R, Berney B, Espinosa-Garcia FJ 1997 Phytoalexins, resistance traits, and domestication status in *Phaseolus coccineus* and *Phaseolus lunatus*. J Chem Ecol 23:1997–2011

Mansfield J 1999 Antimicrobial compounds and resistance: the role of phytoalexins and phytoanticipins. In: Slusarenko AJ, Fraser RSS, VanLoon LC (eds) Mechanisms of resistance to plant diseases. Kluwer Academic, Dordrecht, The Netherlands, in press

Mayama S, Tani T 1982 Microspectro-photometric analysis of the location of avenalumin accumulation in oat leaves in response to fungal infection with *Puccinia coronata*. Physiol Plant Pathol 21:141–149

Muller K, Borger H 1940 Experimentelle Untersuchungen uber die Phytophthora infestans-Resistenz der Kartoffel. Arb Biol Reichsanst Land Forstwirtsch 23:189–231

Nicholson RL, Hammerschmidt R 1992 Phenolic compounds and their role in disease resistance. Annu Rev Phytopathol 30:369–389

Nicholson RL, Kollipara SS, Vincent JR, Lyons PC, Cadena-Gomez G 1987 Phytoalexin synthesis by the sorghum mesocotyl in response to infection by pathogenic and nonpathogenic fungi. Proc Natl Acad Sci USA 84:5520–5524

Pierce ML, Coover EC, Richardson PE, Scholes VE, Essenberg M 1996 Adequacy of cellular phytoalexin concentrations in hypersensitively responding cotton leaves. Physiol Mol Plant Pathol 48:305–324

Rogers EE, Glazebrook J, Ausubel FM 1996 Mode of action of the *Arabidopsis thaliana* phytoalexin camalexin and its role in *Arabidopsis*–pathogen interactions. Mol Plant-Microbe Interact 9:748–757

Scalla R, Matringe M 1994 Inhibitors of protoporphirynogen oxidase as herbicides: diphenyl ethers and related photobleaching molecules. Rev Weed Sci 6:103–132

Snyder BA, Nicholson RL 1990 Synthesis of phytoalexins in sorghum as a site specific response to fungal ingress. Science 248:1637–1639

Snyder BA, Leite B, Hipkind J, Butler LG, Nicholson RL 1991 Accumulation of sorghum phytoalexins induced by *Colletotrichum graminicola* at the infection site. Physiol Mol Plant Pathol 39:463–470

Thomzik JE, Stenzel K, Stocker R, Schreier PH, Hain R, Stahl DJ 1997 Synthesis of a grapevine phytoalexin in transgenic tomatoes (*Lycopersicon esculentum* Mill.) conditions resistance against *Phytophthora infestans*. Physiol Mol Plant Pathol 51:265–278

Tsuji J, Jackson EP, Gage DA, Hammerschmidt R, Somerville SC 1992 Phytoalexin accumulation in *Arabidopsis thaliana* during the hypersensitive reaction to *Pseudomonas syringae* pv. *syringae*. Plant Physiol 98:1304–1309

Tsuji J, Zook M, Hammerschmidt R, Last R, Somerville S 1993 Evidence that tryptophan is not a direct biosynthetic intermediate of camalexin in *Arabidopsis thaliana*. Physiol Mol Plant Pathol 43:221–229

VanEtten H, Matthews P, Teftmeier K, Dietert MF, Stein JI 1989 Phytoalexin detoxification: importance for pathogenicity and practical implications. Annu Rev Phytopathol 27:143–164

VanEtten H, Funnell-Baerg D, Wasmann C, Mcclusky K 1994 Location of pathogenicity genes on dispensable chromosomes in *Nectria haematococca* MPVI. Antonie van Leeuwenhoek 65:263–267

Wasmann CC, VanEtten H 1996 Transformation-mediated chromosome loss and disruption of a gene for pisatin demethylase decreases the virulence of *Nectria haematocca* on pea. Mol Plant-Microbe Interact 9:793–803

Zhao J, Last RL 1996 Coordinate regulation of the tryptophan biosynthetic pathway and indolic phytoalexin accumulation in *Arabidopsis*. Plant Cell 8:2235–2244

Zhou N, Tootle TL, Tsui F, Klessig DF, Glazebrook J 1998 *PAD4* functions upstream from salicylic acid to control defense responses in *Arabidopsis*. Plant Cell 10:1021–1030

Zook M 1998 Biosynthesis of camalexin from tryptophan pathway intermediates in cell-suspension cultures of *Arabidopsis*. Plant Physiol 118:1389–1393

Zook M, Hammerschmidt R 1997 Origin of the thiazole ring of camalexin, a phytoalexin from *Arabidopsis thaliana*. Plant Physiol 113:463–468

Zook M, Hohn T, Bonnen AM, Tsuji J, Hammerschmidt R 1996a Characterization of novel sesquiterpenoid biosynthesis in tobacco expressing a fungal sesquiterpene cyclase. Plant Physiol 112:311–318

Zook M, Johnson K, Hohn T, Hammerschmidt R 1996b Structural characterization of 15-hydroxytrichodiene, a novel sesquiterpenoid produced by transformed tobacco cell suspension cultures expressing a trichodiene synthase gene from *Fusarium sporotrichiodes*. Phytochemistry 43:1235–1237

Zook M, Leege, L, Jacobson D, Hammerschmidt R 1998 Camalexin accumulation in *Arabis lyrata*. Phytochemistry 49:2287–2289

DISCUSSION

Boland: The biosynthesis of many of the terpenoid phytoalexins is linked to the octadecanoid pathway. Do any of the inducing herbicides turn on octadecanoid signalling?

Hammerschmidt: Not that I know of. There is another herbicide, acifluorofen, which has the same mode of action as lactofen. This compound has been shown to induce isoflavonoid accumulation in soybean (Cosio et al 1985). We have not yet looked at soybean beyond the observations reported here.

Scheel: Some work by Ebel and co-workers (Mithofer et al 1997) suggested that in cultured soybean cells the reactive oxygen species are not involved in inducing phytoalexin synthesis. You were suggesting that reactive oxygen species generated by herbicide action are involved in stimulating phytoalexin production. Is there any experimental evidence for this?

Hammerschmidt: No. This is based on the mode of action of these herbicides. These compounds cause a fairly high accumulation of protoporphyrin which leaks out of the plastids and mitochondria, and associates with the plasma membrane. In the light, the protoporphyrin produces active oxygen species and you start to see membrane disruption. The herbicides thus cause the leaves to begin to desiccate. Soybean leaves are slightly damaged by these herbicides. The paper you mention is interesting because it would suggest that this is working by a different mechanism than the way it's actually working as a herbicide — or it could just be the difference between a whole plant and cell suspensions. We haven't been able to show any systemic effects, which is different from what we would expect from a classical system acquired resistance (SAR)-type response.

Pickett: The study of the effect of selective herbicides on 'non-susceptible' plants is an area of great activity. They can give surprising effects, and this is most interesting.

Hammerschmidt: The plants show a certain amount of stress, but they can outgrow it very quickly. The resistance also lasts for only a short period. If we come back a few weeks later we find no more elevated levels of glyceollin. We think the reason that this is working to control the disease is that we are able to time the applications to when *Sclerotinia* is producing ascospores.

Dietrich: Have you measured salicylate levels in any of your plants?

Hammerschmidt: No.

Scheel: I have a question concerning the possible role of phytoalexins in SAR. Has anyone studied the SAR responses in phytoalexin-minus mutants of *Arabidopsis*?

Hammerschmidt: I don't know whether anyone has done that. Zhou et al (1998) have published that they can enhance the ability of one phytoalexin-deficient mutant to produce camalexin with salicylate treatment.

Schultz: If we are going to believe in phytoalexins, you need to settle this issue with a knockout. My recollection is that the Dixon group is close to having the genes for the last steps in one or more of these pathways.

Lamb: Yes, this is in alfalfa.

Schultz: When my students — who are fundamentally ecologists who realize that to get a job they're going to need to learn molecular techniques — are looking for projects to test the adaptive significance of biochemistry, I tell them that they need to be able to knock something out, but only do one thing. This system comes to mind, because it seems to me that we're getting close to being able to just stop the last product.

Hammerschmidt: This is why I think a system like this is potentially cleaner than some of the legume systems, which have been studied a lot more extensively. Although we know a lot more about the role of phytoalexins in beans and alfalfa, the pathways there are more complicated.

Gershenzon: One of the problems in really understanding what phytoalexins do is their mode of action. One doesn't hear much on this topic. Is there anything known about the mode of action of camalexin?

Hammerschmidt: There was a paper from Rogers et al (1996), but they did not have a strong conclusion as to its mode of action. It does affect some permeability functions in the bacterium they tested, but it does not seem to be a site-specific mode of action. The isoflavonoid phytoalexins can also disrupt membranes, but it is still not a very specific mode of action.

It would be nice to know the mechanism because then we could develop a line of fungus or bacterium which could tolerate high levels of the phytoalexin without changing anything else in the pathogen. This would allow us to determine whether phytoalexin tolerance makes any difference in the host–pathogen interaction. I think that the studies of VanEtten with *Nectria* on pea (described in my paper) are illuminating because if you disrupt the gene responsible for phytoalexin detoxification (*pda*) (Wasmann & VanEtten 1996) while keeping the rest of the pathogenicity genes, you find that virulence is decreased by a certain amount. In contrast, if you put the *pda* gene into a pathogen of corn, the corn pathogen is able to cause a larger lesion on pea plants. It does not become a full blown pathogen, but it also doesn't cause the same restricted lesion. This tells us that the phytoalexins are playing an incremental role in the process that stops the pathogen. If you put one of these genes into a saprophyte, it makes no difference whatsoever because the saprophytes do not have the machinery that allows them to attack the plant.

Pickett: Have you spoken confidentially to any companies about camalexin in terms of its mode of action?

Hammerschmidt: No.

Pickett: The companies probably hold some unpublished knowledge on this.

Hammerschmidt: Camalexin has been reported to have some antitumour activities. It has some structural relatedness to the fungicide thiabendazole, which is a mitotic inhibitor. This could explain why this compound might have some effect on tumour development. We haven't looked to see whether or not camalexin inhibits mitotic function in the fungi.

Poppy: Speaking as an ecologist, one of the things that we've thought about doing with our bean system was challenging the beans with a pathogen, and then introducing aphids onto it to see what happens. Have any of the people here working on tritrophic systems including parasitoids looked at plants that have been challenged with pathogens and subsequently exposed to herbivores, to see whether there's an increase in the production of volatiles and subsequent changes in parasitoid foraging behaviour? Has the system been 'primed' at all by the pathogens?

Städler: We tested several phytoalexins and found that the carrot fly laid more eggs on induced leaves (Stanjek et al 1997). We can speculate that some of the bacteria known to induce these phytoalexins are associated with some of the flies living on the plants.

Pickett: As an aside, cattle have lower weight gain when eating meal from oil seed rape (*Brassica napus*) that has been lightly infected with stem canker, because this induces glucosinolates which are a negative nutritional component to cattle.

References

Cosio EG, Weissenbock G, McClure JC 1985 Acifluorofen induced soybean isoflavonoids and enzymes of their biosynthesis in mature soybean leaves: whole leaf and mesophyll responses. Plant Physiol 78:14–19

Mithofer A, Daxberger A, Fromholdtreu D, Ebel J 1997 Involvement of an NADPH oxidase in the elicitor-inducible oxidative burst of soy bean. Phytochemistry 45:1101–1107

Rogers EE, Glazebrook J, Ausubel FN 1996 Mode of action of the *Arabidopsis thaliana* phytoalexin camalexin and its role in *Arabidopsis*–pathogen interactions. Mol Plant-Microbe Interact 9:748–757

Stanjek V, Herhaus C, Ritgen U, Boland W, Städler E 1997 Changes in the leaf surface chemistry of *Apium graveolens* (Apiaceae) stimulated by jasmonic acid and perceived by a specialist insect. Helv Chim Acta 80:1408–1420

Wasmann CC, VanEtten H 1996 Transformation-mediated chromosome loss and disruption of a gene for pisatin demethylase decreases the virulence of *Nectria haematococca* on pea. Mol Plant-Microbe Interact 9:793–803

Zhou N, Tootle TL, Tsui F, Klessig DF, Glazebrook J 1998 *PAD4* functions upstream from salicylic acid to control defense responses in *Arabidopsis*. Plant Cell 10:1021–1030

Octadecanoid and hexadecanoid signalling in plant defence

Elmar W. Weiler, Dietmar Laudert, Boguslawa A. Stelmach, Peter Hennig, Christian Biesgen and Ines Kubigsteltig

Lehrstuhl für Pflanzenphysiologie, Ruhr-Universität, D-44780 Bochum, Germany

Abstract. Plants respond to situations requiring the initiation of inducible defence reactions with a complex array of signalling events that ultimately result in the activation of sets of defence genes. Among the chemical signals involved in the induction of defence reactions are cyclic oxylipins derived from C_{18}- or C_{16}- unsaturated fatty acids, the octadecanoids and the hexadecanoids. Key to understanding octadecanoid biology are the C_{18}-metabolite 12-oxophytodienoic acid (OPDA) and the C_{12}-compound jasmonic acid which is biosynthetically derived from 12-oxophytodienoic acid. Different octadecanoids likely have different biological functions. The bouquet of signalling compounds, rather than any single compound, is probably decisive for the biological response that results. This means that the processes regulating the pool sizes of different octadecanoids and their distribution within the plant are key to understanding octadecanoid biology. Recent results, including the cloning of several enzymes of the octadecanoid biosynthetic pathway, have provided first insights into these processes and into how the octadecanoid system is linked to other defence-related signalling pathways of the plant cell.

1999 Insect–plant interactions and induced plant defence. Wiley, Chichester (Novartis Foundation Symposium 223) p 191–204

Biology of octadecanoids

Since several aspects of octadecanoid biology are dealt with in depth in other chapters, only a brief synopsis will be given here. The physiological processes involving octadecanoids can be divided into two major classes of responses: (a) defence responses against herbivore and pathogen attacks, and (b) processes involving growth inhibition and the promotion of senescence. There is considerable overlap in octadecanoid-controlled biochemical reactions in either of the response classes, which points to a common evolutionary basis (for review see Weiler et al 1998). The view that jasmonic acid (JA) was the sole biologically active octadecanoid had to be abandoned when it was shown that 12-oxophytodienoic acid (OPDA) had biological activity *per se* (Weiler et al 1994), that different physiological processes had different structural requirements for an octadecanoid to be active (Blechert et al 1997), that OPDA vs. JA responses could be separated by appropriate molecular design (Blechert et al 1997, 1999), and that some physiological processes, such as mechanotransduction in *Bryonia dioica* and *Phaseolus vulgaris* internodes (Stelmach et al 1998) or the elicitation of secondary product formation in plant tissue cultures (Parchmann et al 1997) proceeded with transient changes of endogenous levels of OPDA, but not of JA or involved complex patterns of changes in octadecanoid profiles over time. In addition, hexadecanoids may further increase the complexity of oxylipin signatures (Weber et al 1997), and the application of exogenous JA to a tissue may bring about drastic alterations in the pool sizes of other octadecanoids, due to the feedback and feedforward mechanisms involved in the regulation of octadecanoid biosynthesis (Laudert & Weiler 1998, see below). Consequently, a very detailed knowledge of the octadecanoid status of a tissue and of the processes regulating it is required before one can hope to fully understand the role octadecanoids play in any given process. Exquisite analytical tools have been developed that now allow the identification and quantitation of many octadecanoids from small amounts of tissue, including analysis of their isomeric (enantiomeric) status (e.g. Okamoto & Nakazawa 1992, Mueller & Brodschelm 1994, Laudert et al 1997), and heavy-isotope-labelled internal standards of the naturally occurring isomers are accessible on a preparative scale through the use of coupled enzymic reactions involving the cloned biosynthetic enzymes (Laudert et al 1997, Schaller & Weiler 1997a,b, Stelmach et al 1998).

Octadecanoids in plant defence

In pioneering papers, Farmer & Ryan (1990) showed that JA is a signal transducer in herbivore defence and Gundlach et al (1992) demonstrated an involvement of JA in the induction of low M_r secondary metabolites, some or most of which may be defence-related. The crucial role of octadecanoids in herbivore defence is backed by several lines of evidence: (a) the *defenseless* mutant (*def*1) of tomato lacks the ability to raise its JA level after wounding, produces little proteinase inhibitor polypeptide and is more susceptible than wild-type tomato to herbivore attacks (Howe et al 1996); (b) the α-linolenic acid-deficient triple mutant of *Arabidopsis thaliana* (*fad*3–2 *fad*7–2 *fad*8) (McConn et al 1997) likewise lacks the ability to accumulate JA after wounding and is highly susceptible to attack by chewing insects; and (c) the JA-insensitive *coi*1 mutant of *A. thaliana* is more susceptible than wild-type plants to herbivore attack (Rojo et al 1998). Although these results prove the crucial involvement of octadecanoids, and of JA in particular, in mediating wound responses, octadecanoids are not the sole players, as JA-independent signalling pathways have recently been shown to exist (e.g. Titarenko et al 1997). Those appear to operate in the immediate vicinity of the wound sites, while the JA-dependent responses are activated throughout the aerial parts of the plant (*A. thaliana*) (Titarenko et al 1997). Radiolabel from applied [^{14}C]JA has been shown to be translocated from shoot to root in tobacco and, in the root, JA is held responsible for inducing the production of the insecticidal nicotine as a reaction to herbivore damage of the shoot (Zhang & Baldwin 1997). However, in a recent study using transgenic *A. thaliana* plants expressing β-glucuronidase under the control of an octadecanoid- as well as wound-responsive promoter, Kubigsteltig et al (1999) have shown that the signals travelling from a wound site in a leaf to systemic tissues of the shoot are not octadecanoids and that, *vice versa*, applied octadecanoids (OPDA, JA) as well as the octadecanoid analogue, coronatine, act in a strictly local manner and are not transported out of the leaf where they were applied. Thus, it appears that octadecanoids are local mediators of defence reactions. They probably act inside the cell where they are produced (Fig. 1). Signals inducing JA accumulation distant from the primary wound site can be chemical in nature (e.g. the octadecapeptide systemin in tomato (for review see Schaller & Ryan 1996), or physical. Among the physical signals, hydraulic waves (for review see Malone & Alarcon 1995) or, rather, electrical signals (Wildon et al 1992) have been discussed. This issue is not yet solved definitely. Initially, local JA production at the wound site in all probability is a result of decompartmentation in the damaged areas which liberates fatty acids through hydrolysis of lipids. These are then available for conversion to JA. JA production in an undamaged cell may follow a different pathway.

The role(s) of octadecanoids in pathogen defence are understood less well (for review see Dong 1998, Weiler et al 1998). It is beyond doubt that both in cell cultures as well as in differentiated plants, octadecanoids induce the accumulation of a broad spectrum of secondary metabolites (Gundlach et al 1992) such as alkaloids (e.g. macarpine, nicotine), glucosinolates and phenolics (for review see Weiler et al 1998). Many of these have been shown to be either toxic to insects (nicotine) or to microorganisms (macarpine), or are deterrents (glucosinolates). The ecophysiological roles of secondary plant metabolites are well documented and frequently involve microbial defence (Harborne 1988). Plant phenolics further serve as extracellular, unspecific chemicals that denature proteins (e.g. those of invading microorganisms) and that are polymerized to form macromolecules which are not easily degraded, such as tannins and lignin. Such chemical and structural barriers help the plant to fight invaders. Induced lignification as a result of JA application or endogenous production of octadecanoids has been demonstrated (Kaiser et al 1994). Inducible chemical defence against microorganisms (phytoalexins) is usually restricted to sites of infection and is not systemic, and local production of octadecanoids as a consequence of the (likewise local) accumulation of elicitors may trigger this

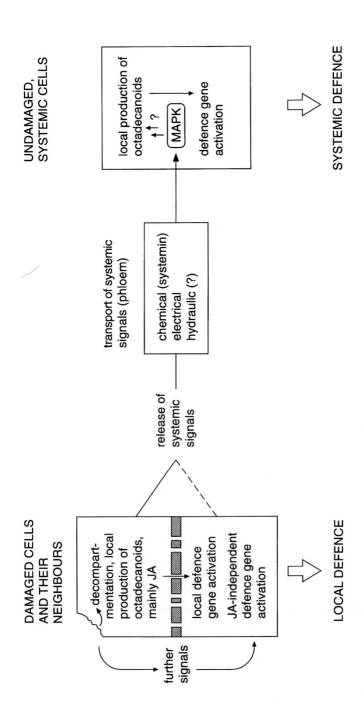

FIG. 1. Local and systemic signalling in wounding-induced defence gene activation. MAPK, mitogen activated protein kinase-like protein kinase.

process. In line with this is the observation that fatty acid ω-3 desaturase is rapidly and transiently increased around fungal infection sites in *Petroselinum crispum* (Kirsch et al 1997). The enzyme was hypothesized to provide additional precursor for the production of octadecanoids.

On the other hand, it has been reported that only a subset of defence genes activated by elicitors are at the same time JA-responsive (Rickauer et al 1997), that the defence reactions of barley against *Erysiphe graminis* do not involve JA (Hause et al 1997) and that none of the pathogenesis-related proteins is activated upon wounding or after JA treatment of *A. thaliana* (for review see Wasternack & Parthier 1997). It is thus clear that, as in the case of wound responses, there are octadecanoid-dependent and octadecanoid-independent signalling pathways leading to different sets of defence genes. A first systematic overview of these pathways, based on genetic evidence largely from *A. thaliana*, has been compiled by Dong (1998) (Fig. 2). According to this working model, avirulent pathogens activate pathogenesis-related genes through a pathway involving the production of reactive oxygen species (leading to a hypersensitive response) and salicylic acid (SA) as a signal transducer. This leads to systemic acquired resistance (SAR) characteristic, e.g. for resistance of *A. thaliana* against *Peronospora parasitica* or *Pseudomonas syringae* pv. *maculicola*. Virulent pathogens may also induce SAR through SA, but independently of a hypersensitive response. Other resistance-related genes, such as PDF1.2 and Thi2.1, are induced as a consequence of attack by an avirulent microorganism, by wounding or elicitation through a SA-independent but octadecanoid- and ethylene-dependent signalling pathway, similar to the one operating during a wound response. Resistance to *P. parasitica* involves this octadecanoid/ethylene-dependent pathway. The induced systemic resistance (ISR) response against root-colonizing bacteria is SA-independent but octadecanoid/ethylene-dependent also (for review see Dong 1998). SA represses the octadecanoid/ethylene-dependent pathway and also the reverse may be true (Dong 1998), allowing for fine-tuning of the individual responses. It is thus becoming clear that an important facet of not only plant herbivore defence, but also of plant defences against pathogens, is dependent on the octadecanoid signalling system.

Octadecanoid biosynthesis and its compartmentation

It is generally assumed that in all plants, octadecanoids are produced from α-linolenic acid through the Vick–Zimmerman pathway (Vick & Zimmerman 1984, for a recent review see Mueller 1997). An alternative pathway leads to JA production via dinor-OPDA from hexadeca-7(Z),10(Z),13(Z)-trienoic acid, which is a major unsaturated fatty acid in 16:3 plants such as *A. thaliana* (Weber et al 1997) (Fig. 3). The existence of a pathway starting from linolenic acid and

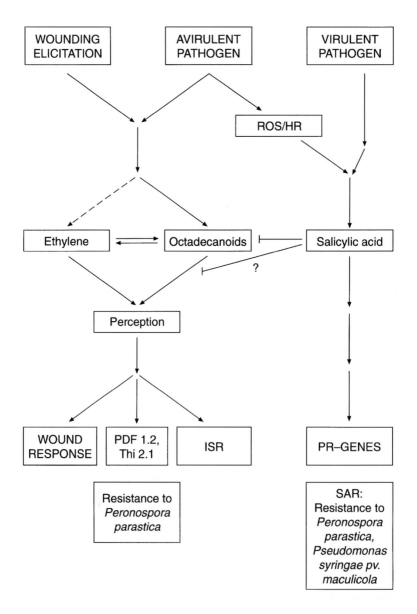

FIG. 2.   Signalling involved in the development of broad-spectrum disease resistance in plants. Adopted from Dong (1998), largely based on genetic evidence from *Arabidopsis thaliana*. ROS, reactive oxygen species; HR, hypersensitive response; PR genes, pathogenesis-related genes; SAR, systemic acquired resistance; ISR, induced systemic resistance; Thi2.1, thionin 2.1; PDF1.2, plant defensin 1.2.

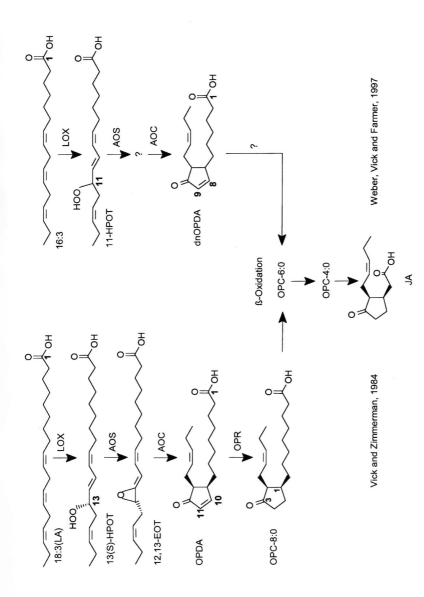

FIG. 3. Biosynthesis of cyclopentanoic oxylipins from α-linolenic acid (LA) or hexadeca-7(Z),10(Z),13(Z)-trienoic acid. LOX, lipoxygenase; AOS, allene oxide synthase; AOC, allene oxide cyclase; OPR, OPDA reductase; 13(S)-HPOT, 13(S)-hydroperoxyoctadecatrienoic acid; 12,13-EOT, 12,13-epoxyoctadecatrienoic acid; OPDA, 12-oxophytodienoic acid; OPC-8(6,4):0, 3-oxo-2(2'(Z)-pentenyl)cyclopentane-1-octanoic (hexanoic, butanoic) acid; JA, jasmonic acid; dnOPDA, dinor-OPDA; 11-HPOT, 11-hydroperoxyhexadecatrienoic acid.

leading to dihydrojasmonic acid (dh-JA) has been proposed based on the low substrate specificity of some of the enzymes involved in JA biosynthesis (Laudert et al 1996, Gundlach & Zenk 1998), but its *in vivo* significance remains in doubt, since it has been shown that allene oxide cyclase (AOC) is highly specific for 12,13-epoxylinolenic acid and does not accept 12,13-epoxylinoleic acid as substrate (Ziegler et al 1997). Furthermore, dh-JA seems restricted to only a few species (e.g. Gundlach & Zenk 1998).

Several enzymes of the biosynthetic pathway have recently been cloned: lipoxygenase (LOX2) (Bell et al 1995), allene oxide synthase (AOS) (Song et al 1993, Pan et al 1995, Laudert et al 1996) and OPDA reductase (OPR) (Schaller & Weiler 1997b), and AOC has been purified to homogeneity (Ziegler et al 1997) and awaits cloning. These, as well as biochemical studies (Blée & Joyard 1996, for review see Maréchal et al 1997) have shed light on the compartmentation of the pathway (Fig. 4). The steps from α-linolenic acid to OPDA are all plastidic and, in leaves, localized in the chloroplast, most likely at the envelope membranes (Blée & Joyard 1996). The intracellular location of AOC is uncertain. The instability of the allene oxide, 12,13-epoxylinolenic acid, in aqueous solution (half-life less than 15–30 s) would favour, as shown in Fig. 4, a plastidic localization of AOC and a close interaction of the two enzymes. However, no direct evidence for this is currently available. OPR is a cytosolic enzyme (Schaller & Weiler 1997a), and the synthesis of JA from OPC-8:0 should occur in microbodies (peroxisomes in the leaves of green plants) as this is the sole compartment of fatty acid β-oxidation in the plant cell.

Any local or systemic signal activating octadecanoid biosynthesis must consequently transverse the cell to reach the chloroplast, and effective means of metabolite transport from the chloroplast to the cytosol and from the cytosol to the peroxisome and vice versa must be established. Nothing is known yet about these processes.

Regulation of octadecanoid biosynthesis

All enzymes of octadecanoid biosynthesis are present in uninduced tissues. Thus, substrate availability will normally limit metabolite flow in the pathway. Results by Harms et al (1995) seem to be at variance with this hypothesis. These authors have found higher resting levels of JA in transgenic potato overexpressing flax AOS. However, no concurrent induction of proteinase inhibitors has been observed. On the contrary, overexpression of the *A. thaliana* AOS in tobacco did not alter the resting level of JA (D. Laudert & E. W. Weiler, unpublished work 1997), in agreement with the notion that substrate availability is indeed limiting JA production in uninduced *A. thaliana* plants (provided that no other limiting process downstream of AOS occurs in this species). It is held that free α-linolenic

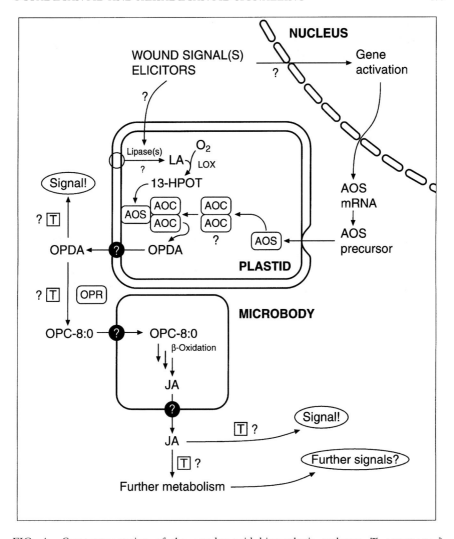

FIG. 4. Compartmentation of the octadecanoid biosynthetic pathway. T, transport; ?, hypothetical; all further abbreviations as in Fig. 3. The intracellular localization of AOC is still uncertain.

acid is the starting substrate for JA production. In chloroplasts, where α-linolenic acid is not being synthesized *de novo*, but is acquired from the endoplasmic reticulum in the form of membrane lipid precursors such as diacylglycerols, free α-linolenic acid consequently should arise from hydrolytic cleavage of membrane lipids by A2-type lipases or acylesterases. The release reaction could thus be an important regulatory step in octadecanoid production, and increased levels of α-linolenic acid as a consequence of elicitation (Mueller et al 1993) or wounding (Conconi et al 1996) have indeed been reported.

Any of the biosynthetic enzymes could be sites of regulation of pathway capacity which should be low in uninduced plants, but high when demand for octadecanoids arises. Finally, release of OPDA from the chloroplast and of JA from the peroxisome must be considered potential points of control. Indeed, there is now evidence for multiple control points in the octadecanoid pathway (Fig. 5). Key features of the regulatory network that is beginning to emerge are the following (Laudert & Weiler 1998):

- AOS is a focal point of pathway control. The enzyme is up-regulated by its own reaction product, OPDA, as well as by the pathway's end-product, JA. This establishes a feedforward amplification that will maximize output from α-linolenic acid to OPDA, since AOS catalyses the limiting step of at least this segment of the pathway. The enzyme is further up-regulated by local as well as systemic wound signals other than octadecanoids.
- Octadecanoids, such as JA, induce ethylene production and ethylene, in turn, induces an increase in AOS activity, so that these two factors are mutually agonistic in stimulating their synthesis.
- Together with the stimulatory effects of ethylene on lipases and of wound factors on lipases and lipoxygenases (the latter are also induced by JA), a network of controls is operative designed to maximally amplify the pathway's capacity and output, once primed. Primary inducers could be wound factors (e. g. systemin), octadecanoids produced in wounds as a result of decompartmentation or elicitors, and probably other factors.

Peña-Cortés et al (1993) have interpreted their findings that SA (and aspirin) inhibited wound-induced accumulation of proteinase inhibitors as well as wound-induced JA in tomato leaves—an inhibition that could be overcome by OPDA or JA, but not by α-linolenic acid or its 13-hydroperoxide—to mean that SA would inhibit a step in octadecanoid biosynthesis prior to the formation of OPDA. However, at least in *A. thaliana*, this does not seem to be the case. SA induced an accumulation of AOS mRNA, polypeptide and enzymic activity with the result that OPDA levels, but not JA levels rose steadily in the treated plants (Laudert & Weiler 1998). Likewise, SA does not inhibit AOS activity (D. Laudert, unpublished work 1997), nor does SA inhibit AOC activity (Ziegler et al 1997). Rather, it seems that SA blocks OPDA-release from chloroplasts (Laudert & Weiler 1998), and this may form the molecular basis for the inhibition of octadecanoid-dependent signalling by SA (see Fig. 2). These results call for a specific mechanism of OPDA release (not simple diffusions from the organelle) and imply the capability of the chloroplast to

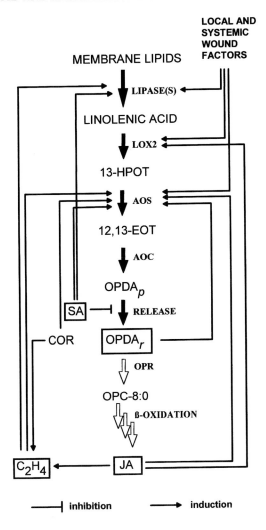

FIG. 5. Regulation of octadecanoid biosynthesis. The compilation is based on the work of Laudert & Weiler (1998) and on others cited and discussed in their paper. SA, salicylic acid; COR, coronatine; OPDA$_p$, plastidic OPDA; OPDA$_r$, OPDA released from the plastid and available to OPR; all other abbreviations as in Fig. 3.

store this JA precursor until the demand for it arises. While at present nothing is known about the mechanisms underlying octadecanoid storage and release, they will be instrumental for a cell to control octadecanoid profiles and, by this, the ensuing cellular responses.

Open questions in octadecanoid signalling

Although progress in the field of octadecanoid biology has been significant and will likely continue to be fast, many open questions remain:

- Which immediate signal triggers the release of α-linolenic acid from membrane lipids, and which is the enzyme responsible for the release process?
- How does OPDA release by chloroplasts occur and how is this process regulated?
- What determines the fate of OPDA as a signal transducer in its own right and/or as a precursor to JA?
- Are there alternative routes for OPDA metabolism?
- Where in the cell are the OPDA and JA receptors localized, what is their structure and how do they act?
- What, if any, components transduce the signal from an octadecanoid receptor to the octadecanoid-responsive genes?
- Are there different sets of OPDA- and JA-responsive genes?

These are only a few of the most pressing questions that need an answer urgently.

Acknowledgements

Work in the author's laboratory was funded by the Deutsche Forschungsgemeinschaft, Bonn, Fonds der Chemischen Industrie, Frankfurt (literature provision) as well as by BMBF, Bonn.

References

Bell E, Creelman RA, Mullet JE 1995 A chloroplast lipoxygenase is required for wound-induced jasmonic acid accumulation in *Arabidopsis*. Proc Natl Acad Sci USA 92:8675–8679

Blechert S, Bockelmann C, Brümmer O et al 1997 Structural separation of biological activities of jasmonates and related compounds. J Chem Soc Perkin Trans I 23:3549–3560

Blechert S, Bockelmann C, Füsslein M et al 1999 Structure–activity analyses reveal the existence of two separate groups of active octadecanoids in elicitation of the tendril coiling response of *Bryonia dioica* Jacq. Planta 207:470–479

Blée E, Joyard J 1996 Envelope membranes from spinach chloroplasts are a site of metabolism of fatty acid hydroperoxides. Plant Physiol 110:445–454

Conconi A, Miquel M, Browse JA, Ryan CA 1996 Intracellular levels of free linolenic and linoleic acids increase in tomato leaves in response to wounding. Plant Physiol 111:797–803

Dong X 1998 SA, JA, ethylene, and disease resistance in plants. Curr Opin Plant Biol 1:316–323

Farmer EE, Ryan CA 1990 Interplant communication: airborne methyl jasmonate induces synthesis of proteinase inhibitors in plant leaves. Proc Natl Acad Sci USA 87:7713–7716

Gundlach H, Zenk MH 1998 Biological activity and biosynthesis of pentacyclic oxylipins: the linoleic acid pathway. Phytochemistry 47:527–537

Gundlach H, Müller MJ, Kutchan TM, Zenk MH 1992 Jasmonic acid is a signal transducer in elicitor-induced plant cell cultures. Proc Natl Acad Sci USA 89:2389–2393

Harborne J 1988 Introduction to ecological biochemistry, 3rd edn. Academic Press, London

Harms K, Atzorn R, Brash A et al 1995 Expression of a flax allene oxide synthase cDNA leads to increased endogenous jasmonic acid (JA) levels in transgenic potato plants but not to a corresponding activation of JA-responding genes. Plant Cell 7:1645–1654

Hause B, Kogel KH, Parthier B, Wasternack C 1997 In barley leaf cells, jasmonates do not act as a signal during compatible or incompatible interactions with the powdery mildew fungus (*Erysiphe graminis* f sp. *hordei*). J Plant Physiol 150:127–132

Howe GA, Lightner J, Browse J, Ryan CA 1996 An octadecanoid pathway mutant (JL5) of tomato is compromised in signaling for defense against insect attack. Plant Cell 8:2067–2077

Kaiser I, Engelberth J, Groth B, Weiler EW 1994 Touch- and methyl-jasmonate-induced lignification in tendrils of *Bryonia dioica* Jacq. Bot Acta 107:24–29

Kirsch C, Takamiya-Wik M, Reinold S, Hahlbrock K, Somssich IE 1997 Rapid, transient, and highly localized induction of plastidial ω-3 fatty acid desaturase mRNA at fungal infection sites in *Petroselinum crispum*. Proc Natl Acad Sci USA 94:2079–2084

Kubigsteltig J, Laudert D, Weiler EW 1999 Structure and regulation of the *Arabidopsis thaliana* allene oxide synthase gene. Planta, in press

Laudert D, Weiler EW 1998 Allene oxide synthase: a major control point in *Arabidopsis thaliana* octadecanoid signalling. Plant J 15:675–684

Laudert D, Pfannschmidt U, Lottspeich F, Holländer-Czytko H, Weiler EW 1996 Cloning, molecular and functional characterization of *Arabidopsis thaliana* allene oxide synthase (CYP 74), the first enzyme of the octadecanoid pathway to jasmonates. Plant Mol Biol 31:323–335

Laudert D, Hennig P, Stelmach BA, Müller A, Andert L, Weiler EW 1997 Analysis of 12-oxo-phytodienoic acid enantiomers in biological samples by capillary gas chromatography-mass spectrometry using cyclodextrin stationary phases. Anal Biochem 246:211–217

Malone M, Alarcon J-J 1995 Only xylem-borne factors can account for systemic wound signalling in the tomato plant. Planta 196:740–746

Maréchal E, Block MA, Dorne AJ, Douce R, Joyard J 1997 Lipid synthesis and metabolism in the plastid envelope. Physiol Plant 100:65–77

McConn M, Creelman RA, Bell E, Mullet JE, Browse J 1997 Jasmonate is essential for insect defense in *Arabidopsis*. Proc Natl Acad Sci USA 94:5473–5477

Mueller MJ 1997 Enzymes involved in jasmonic acid biosynthesis. Physiol Plant 100:653–663

Mueller MJ, Brodschelm W 1994 Quantification of jasmonic acid by capillary gas chromatography-negative chemical ionization-mass spectrometry. Anal Biochem 218:425–435

Mueller MJ, Brodschelm W, Spannagl E, Zenk MH 1993 Signaling in the elicitation process is mediated through the octadecanoid pathway leading to jasmonic acid. Proc Natl Acad Sci USA 90:7490–7494

Okamoto M, Nakazawa H 1992 Direct chromatographic separation of the enantiomers of methyl jasmonate and its derivatives. Biosci Biotech Biochem 56:1172–1173

Pan Z, Durst F, Werck-Reichhart D et al 1995 The major protein of guayule rubber particles is a cytochrome P_{450}: characterization based on cDNA cloning and spectroscopic analysis of the solubilized enzyme and its reaction products. J Biol Chem 270:8487–8494

Parchmann S, Gundlach H, Mueller MJ 1997 Induction of 12-oxo-phytodienoic acid in wounded plants and elicited plant cell cultures. Plant Physiol 115:1057–1064

Peña-Cortés H, Albrecht T, Prat S, Weiler EW, Willmitzer L 1993 Aspirin prevents wound-induced gene expression in tomato leaves by blocking jasmonic acid biosynthesis. Planta 191:123–128

Rickauer M, Brodschelm W, Bottin A, Veronesi C, Grimal H, Esquerré-Tugayé MT 1997 The jasmonate pathway is involved differentially in the regulation of different defense responses in tobacco cells. Planta 202:155–162

Rojo E, Titarenko E, León J, Berger S, Vancanneyt G, Sánchez-Serrano JJ 1998 Reversible protein phosphorylation regulates jasmonic acid-dependent and acid-independent wound signal transduction pathways in *Arabidopsis thaliana*. Plant J 13:153–165

Schaller A, Ryan CA 1996 Systemin — a polypeptide defense signal in plants. BioEssays 18: 27–33

Schaller F, Weiler EW 1997a Enzymes of octadecanoid biosynthesis in plants: 12-oxo-phytodienoate-10,11-reductase. Eur J Biochem 245:294–299

Schaller F, Weiler EW 1997b Molecular cloning and characterization of 12-oxophytodienoate reductase, an enzyme of the octadecanoid signaling pathway from *Arabidopsis thaliana*. Structural and functional relationship to yeast old yellow enzyme. J Biol Chem 272:28066–28072

Song W-C, Funk CD, Brash AR 1993 Molecular cloning of an allene oxide synthase: a cytochrome P_{450} specialized for the metabolism of fatty acid hydroperoxides. Proc Natl Acad Sci USA 90:8519–8523

Stelmach BA, Müller A, Hennig P, Laudert D, Andert L, Weiler EW 1998 Quantitation of the octadecanoid 12-oxo-phytodienoic acid, a signalling compound in mechanotransduction. Phytochemistry 47:539–546

Titarenko E, Rojo E, León J, Sánchez-Serrano JJ 1997 Jasmonic acid-dependent and acid-independent signaling pathways control wound-induced gene activation in *Arabidopsis thaliana*. Plant Physiol 115:817–826

Vick BA, Zimmerman DC 1984 Biosynthesis of jasmonic acid by several plant species. Plant Physiol 75:458–461

Wasternack C, Parthier B 1997 Jasmonate-signalled plant gene expression. Trends Plant Sci 2:302–307

Weber A, Vick BA, Farmer EE 1997 Dinor-oxo-phytodienoic acid: a new hexadecanoid signal in the jasmonate family. Proc Natl Acad Sci USA 94:10473–10478

Weiler EW, Kutchan TM, Gorba T, Brodschelm W, Niesel U, Bublitz F 1994 The *Pseudomonas* phytotoxin coronatine mimics octadecanoid signalling molecules of higher plants. FEBS Lett 345:9–13

Weiler EW, Laudert D, Schaller F, Stelmach BA, Hennig P 1998 Fatty acid-derived signaling molecules in the interaction of plants with their environment. In: Romeo JT, Downum KR, Verpoorte R (eds) Phytochemical signals and plant–microbe interactions. Plenum, New York (Recent Adv Phytochem 32) p 179–205

Wildon DC, Thain JF, Minchin PEH et al 1992 Electrical signalling and systemic proteinase inhibitor induction in the wounded plant. Nature 360:62–65

Zhang Z-P, Baldwin IT 1997 Transport of [2-^{14}C] jasmonic acid from leaves to roots mimics wound-induced changes in endogenous jasmonic acid pools in *Nicotiana sylvestris*. Planta 203:436–441

Ziegler J, Hamberg M, Miersch O, Parthier B 1997 Purification and characterization of allene oxide cyclase from dry corn seeds. Plant Physiol 114:565–573

Induced plant defence responses: scientific and commercial development possibilities

Robert A. Dietrich, Kay Lawton, Leslie Friedrich, Rebecca Cade, Mike Willits and Klaus Maleck

Novartis Agribusiness Biotechnology Research Inc., 3054 Cornwallis Road, Research Triangle Park, NC 27709, USA

Abstract. Recent work has demonstrated that plants have endogenous defence mechanisms that can be induced as a response to attack by insects and pathogens. There are two well-studied examples of these induced defence responses. Systemic acquired resistance (SAR) results in increased resistance to a broad spectrum of pathogens throughout a plant in response to localized necrosis caused by pathogen infection. The second example is the systemic induction of proteinase inhibitors to deter feeding by herbivores following an initial event of feeding. In addition, there is now preliminary evidence for other induced defence response pathways. By understanding the breadth of induced defence responses and the mechanisms used to control these pathways, novel plant protection strategies may be developed for use in agronomic settings. Rather than reducing crop losses caused by pests or pathogens by using chemicals that are designed to kill the offending organism, the plant's own defence mechanisms can be used to limit damage due to pests. Novel crop protection strategies based on genetic or chemical regulation of these induced responses show great potential. The first example of a crop protection product that acts by inducing an endogenous defence response pathway is now on the market. Bion® reduces the level of pathogen infection in plants by activating SAR.

1999 Insect–plant interactions and induced plant defence. Wiley, Chichester (Novartis Foundation Symposium 223) p 205–222

Agriculture is faced with the challenge of feeding an ever-increasing global population on a limited amount of land, making strategies to reduce losses due to pests and pathogens increasingly important. Current crop protection strategies include both chemical approaches, using pesticides which are toxic to the offending organism, and genetic approaches, involving the incorporation of resistance genes into the germplasm of the plant. Both approaches have proven effective, yet both have limitations. Pesticides and resistance genes generally target a single factor in the organism, meaning that a single mutation in the pest is usually sufficient to allow it to overcome resistance. Also, resistance genes are generally very specific, providing resistance against only one particular isolate of one species of pathogen. One way to avoid some of these problems is to develop new strategies for crop protection to supplement current practices. A novel approach is to use a plant's own induced defence responses. By directed expression of specific defence response pathways, it may be possible to significantly enhance resistance to pathogens and pests in an agronomic setting.

Three induced defence response pathways have been identified in plants. Two of these, systemic acquired resistance (SAR) (Ryals et al 1996) and induced systemic resistance (ISR) (Pieterse et al 1996), result in enhanced systemic resistance to some pathogens. The third induced defence response is triggered by wounding and has been studied primarily in the context of induced resistance to insects (Ryan 1990). There is increasing evidence that at least some of the steps in this third pathway may play a role in resistance to pathogens as well. While the wound-induced and ISR pathways offer potential for commercial development, SAR is so far the only induced defence response that has been the basis of a commercial crop protection product. Much of this chapter will focus on SAR as a model for the incorporation of induced defence responses in crop protection strategies.

Systemic acquired resistance

Systemic acquired resistance refers to the activation of pathways resulting in enhanced resistance to infection that develops systemically in a plant following localized necrosis caused by a pathogen (Ryals et al 1996). A basic model of the SAR pathway is shown in Fig. 1A. Cell death at the point of the primary, inducing infection appears to be a key factor required to induce SAR. In an incompatible interaction, this necrosis is the result of an endogenous cell death programme in the plant. Cell death is initiated following the interaction, either direct or indirect, of a specific avirulence gene product in the pathogen and the corresponding resistance gene in the plant, and is referred to as the hypersensitive response (HR). SAR can also be induced in a susceptible plant if the infection results in necrosis. In either case, localized necrosis on a single leaf can result in the establishment of enhanced resistance systemically in the plant. Fungal, bacterial and viral pathogens are all capable of inducing SAR. The systemic resistance that subsequently develops is effective against some (though not all) viral, fungal and bacterial pathogens, no matter what the inducing inoculum is.

The expression of a subset of pathogenesis-related (PR) genes (both locally at the point of infection and systemically) is correlated with the induction of SAR. While overexpression of individual SAR genes in most cases results in only limited enhanced resistance, it is thought that the combined expression of the full set of SAR genes is responsible for the broad spectrum resistance characteristic of SAR.

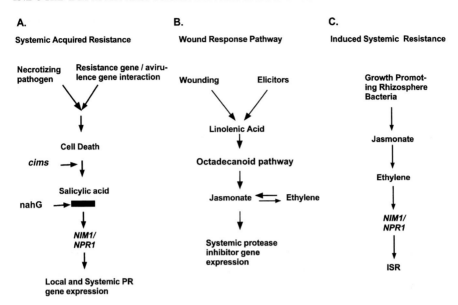

FIG. 1. Basic outlines of three induced plant defence responses. (A) Systemic acquired resistance (SAR). Pathogen-induced cell death leads to accumulation of salicylic acid (SA) which triggers the local and systemic induction of pathogenesis-related genes via the action of the product of the *NIM1/NPR1* gene. The *cim* mutants appear to affect functions between cell death and SA, as they are characterized by constitutively elevated levels of SA and constitutive SAR in the absence of cell death. *nahG* represents a synthetic mutant that is deficient in SA, generated by expressing the bacterial salicylate hydrolase gene in the plant. The *NIM1/NPR1* gene is an essential component of the signal transduction pathway leading from SA to SAR gene expression. (B) Wound response pathway. The wound response pathway is triggered by wounding and fungal elicitors. Based on results from tomato, both jasmonate and ethylene are required for induction of the *pin* genes, and neither alone is sufficient. (C) Induced systemic resistance (ISR). The signal transduction pathway for ISR includes elements from both the SAR and wound-induced pathways. Both jasmonate and ethylene sensitivity are required for ISR as in the wound response pathway, while a functional *NIM1/NPR1* gene is also required. There are no genes as yet identified whose expression correlates with ISR. It should be noted that the SAR results are from a variety of plants, including *Arabidopsis* and tobacco, the wound response results are primarily from tomato, and ISR results are primarily from studies on *Arabidopsis*.

Several lines of evidence point to a key role for salicylic acid (SA) as a signalling molecule in the induction of SAR. First, endogenous SA levels increase just prior to the development of SAR (Malamy et al 1990, Metraux et al 1990, Rasmussen et al 1991). Second, treatment of plants with exogenous SA leads to enhanced disease resistance with all the characteristics of SAR (White 1979). Finally, plants which express a protein that degrades SA to inactive compounds (salicylic acid hydrolase, encoded by the *nahG* gene from the bacterium *Pseudomonas putida*) cannot accumulate SA and are unable to develop SAR (Gaffney et al 1993). Taken together these experiments indicate that SA is both necessary and sufficient for the induction of SAR.

A genetic approach has been very useful in elucidating steps in the SAR pathway. Mutants with phenotypes suggesting defects at various points in SAR signalling have been identified. One class of mutants exhibits constitutive immunity. In these *cim* mutants, the downstream features of SAR, PR gene expression and broad spectrum resistance (Lawton et al 1993) are constitutively expressed. There is no cell death associated with the *cim* phenotype but they all have elevated constitutive levels of SA. Thus, the corresponding genes may represent key regulatory genes in the SAR pathway between cell death and SA. The *cim* mutants can be placed into classes based on the expression patterns of a number of defence-related genes, both SA dependent and SA independent, and on the resistance specificities to a range of pathogens.

To identify genes acting in the SAR pathway between SA and resistance, mutants were selected that no longer showed increased resistance following chemical induction of SAR. A single locus was identified in four independent screens: *non-inducible immunity* (*nim1*) (Delaney et al 1995) *non-inducible PR* (*npr1*) (Cao et al 1994), *salicylic acid insensitive* (*sai1*) (Shah et al 1997), and *enhanced disease susceptibility* (*eds-5*) (Glazebrook et al 1996). The mutants are defective both in the local response to infection and in PR gene expression and the establishment of SAR normally induced following either biological or chemical induction of SAR. The *NIM1/NPR1* gene has been cloned (Cao et al 1997, Ryals et al 1997), and though the exact mechanism of action remains unknown, it clearly plays an important role in the regulation of SAR.

Developing SAR for use in crop protection

With the emerging understanding of the SAR pathways, attempts are being made to incorporate SAR in crop protection strategies. Two approaches are currently being attempted, a chemical approach and a molecular/genetic approach.

As previously noted, SA is an essential component of the SAR pathway, and exogenous application of SA can induce SAR. However, because the concentration of SA required to induce SAR is very close to the concentration that is phytotoxic, SA is not practical for use on a commercial basis. An extensive screen was undertaken to identify other chemicals that could induce SAR as well or better than SA, while reducing the undesirable side effects. Two chemicals were found; 2,6-dichloroisonicotinic acid (INA) (Metraux et al 1991) and benzo(1,2,3)thiadiazole-7-carbothioic acid S-methyl ester (BTH) (Friedrich et al 1996, Gorlach et al 1996, Lawton et al 1996). As shown in Fig. 2, these compounds share structural similarities with SA. They induce expression of the same genes that are induced during SAR, and induce enhanced resistance to the

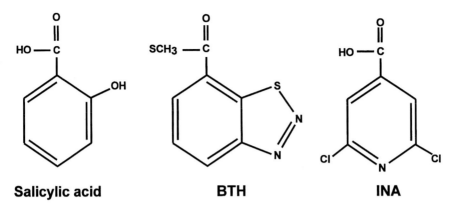

Salicylic acid **BTH** **INA**

FIG. 2. The structures of SAR-inducing compounds. Salicylic acid is the endogenous signalling molecule, while BTH (benzo(1,2,3)thiadiazole-7-carbothioic acid S-methyl ester) and INA (2,6-dichloroisonicotinic acid) are synthetic compounds that effectively induce SAR.

same pathogens as SAR. BTH is the basis of a commercial plant protection compound, Bion®, which is being developed for use worldwide on a variety of crops.

Since mutations in the *NIM1/NPR1* gene result in a severe reduction in the SAR response, experiments were done to see if overexpression of *NIM1/NPR1* would have the opposite effect, resulting in plants with constitutive enhanced resistance. Using the 35S promoter linked to a *NIM1/NPR1* cDNA, transgenic lines were generated that overexpressed the introduced gene to varying levels (Cao et al 1998). A positive correlation between the level of NIM1/NPR1 protein in the transgenic plants and the degree of constitutive resistance to pathogens was observed.

The BTH and *NIM1/NPR1* overexpression results suggest that both chemical and molecular approaches may yield tools for incorporating SAR in disease control programs. An even better strategy may be a combination of the two approaches. For example, while *NIM1/NPR1* overexpression alone can lead to increased resistance in *Arabidopsis*, strong resistance was seen only in the lines with the highest expression of the transgene (Cao et al 1998). A second independent set of experiments confirmed and extended these results. While lines expressing the *NIM1/NPR1* transgene at high levels were constitutively resistant, all transgenic lines showed enhanced response to BTH. Full disease resistance could be activated in transgenic plants at BTH concentrations 10–20-fold lower than that required on untransformed plants. Reflecting the increased sensitivity, the BTH-treated transgenic *NIM1/NPR1* plants exhibited a faster and stronger induction of PR gene expression following treatment. Thus, increased expression of *NIM1/NPR1* makes the plants more responsive to BTH (L. Friedrich & K. Lawton, unpublished data).

Other induced responses

A second well-studied induced defence response in plants is the wound response pathway (Fig. 1B). This pathway has been studied primarily in the context of insect damage (Ryan 1990), and most extensively in tomato. Insect feeding triggers the pathway via intermediates including jasmonic acid (JA) and ethylene (O'Donnell et al 1996). The ultimate result is the expression of a battery of genes, including protease inhibitor (*pin*) genes. The protease inhibitors interfere with insect digestion and discourage further feeding. At least some steps in this pathway may be involved in protection from pathogens as well. The pathway can be triggered by oligosaccharide elicitors produced during some host–pathogen interactions (Gundlach et al 1992). Also, induction of the wound response pathway can induce certain responses that are effective against pathogens, including biosynthesis of antimicrobial phytoalexins and the expression of a subset of PR genes (Choi et al 1994, Xu et al 1994). Finally, mutants that are defective in jasmonate biosynthesis or response show increased susceptibility to certain pathogens (Vijayan et al 1998). Thus, both SAR and the wound response pathway may be effective against pathogens, but the respective pathways control different subsets of PR genes and may be effective against different or possibly partially overlapping sets of pathogens.

A third induced defence response shares some features with both SAR and the wound response pathways. ISR is induced in plants by certain growth-promoting rhizosphere bacteria and is effective against some pathogens (Pieterse et al 1996). ISR signalling requires JA and ethylene (Fig. 1C), and it can be induced in *Arabidopsis* plants that express the *nahG* gene, indicating that ISR signalling is SA independent. In these respects, ISR is similar to the wound response pathway, and unlike SAR. However, ISR is reported to be dependent on a functional *NIM1/NPR1* gene, like SAR (Pieterse et al 1998) but unlike the wound response. Thus, ISR appears to require components of both the SAR and wound response pathways. (Note that genetic data relating to ISR places jasmonate upstream of ethylene in a linear pathway while in the wound response pathway these are not in a linear pathway. This may be due to different signalling pathways for the two responses or because the studies were done in different plants, tomato for the wound response and *Arabidopsis* for ISR.)

Considerations in adapting induced responses for crop protection

If these induced responses are to be developed for use in crop protection, a number of factors must be considered. One of the primary concerns is the potential for pleiotropic effects of altering the regulation of any pathway. This is especially true with the SAR and wound response pathways, as there is evidence for mutual antagonism between them. Exogenous treatment with SA has been shown to inhibit multiple steps in the wound response pathway (Doherty et al 1988, Peña-Cortés et al 1993, Doares et al 1995), resulting in reduced *pin* gene induction following wounding. This suggests that induction of SAR could repress the wound response. Thus, while induction of SAR would make a plant more resistant to pathogens, it could at the same time make the plant more susceptible to insect damage. Experiments to test for interactions between the pathways have given mixed results. Bostock et al found induction of SAR in tomatoes by BTH resulted in a slight increase in survivorship of an arthropod herbivore compared to control plants. Interestingly, induction of SAR by a bacterial pathogen resulted in the induction of both *pin* and PR gene expression, and an increased resistance to both pathogens and insects. This suggests that biological induction of SAR can engage both the SAR- and wound-induced pathways (Fidantsef et al 1999, Stout et al 1999). In independent experiments, induction of SAR in tobacco with tobacco mosaic virus did not result in increased feeding by tobacco hornworm larvae (Ajlan & Potter 1992). There is no evidence that damage caused by insect feeding represses resistance to pathogens. In fact, a number of experiments have found an increase in pathogen resistance following induction by insect feeding (Karban et al 1987, Stout et al 1998).

The potential pleiotropic effects of ectopic expression of these pathways may extend beyond defence response pathways. It is possible that some of the components of these pathways may impinge on other non-defence pathways as well. For example, the jasmonate-insensitive *Arabidopsis* mutant *coi1* and a jasmonate-deficient triple mutant are male sterile, suggesting a possible role for jasmonate in development as well as in defence responses (Feys et al 1994, McConn & Browse 1996).

An important consideration in attempting to use these endogenous pathways in crop protection is the possible metabolic cost of overexpression of induced response pathways. While the cost of diverting resources into the expression of the battery of SAR genes could make the plants less competitive evolutionarily, it is not clear if constitutive expression would have a significant impact on yield in an agricultural setting where resources are less limiting. In field tests, no yield losses have been detected following treatment with Bion®. In a series of experiments looking at the cost of induction of the jasmonate pathway, Baldwin (1998) found a slight reduction in seed yield comparing jasmonate-treated with untreated plants in the absence of herbivorous insects. In plants with intermediate levels of herbivore attack, however, the jasmonate-treated plants produced more seed than the untreated plants. Thus, the cost of the induced defence response is more than compensated for by the reduction in damage from pests.

While the potential for pathogens and pests to overcome resistance is a consideration in any crop protection strategy, it may be a less serious problem when using these endogenous induced responses. Because the resistance is based on the expression of a battery of genes, this resistance should be more durable than resistance based, for example, on a single resistance gene.

An additional benefit of the incorporation of endogenous defence pathways in crop protection may be in a reduction in the amount of pesticide required for effective pest control. Recent evidence suggests that treatment of plants with BTH enhances the efficacy of some fungicides (Molina et al 1998). By reducing the amount of fungicide required, costs are reduced and the effective life of the fungicide may be extended because of reduced selective pressure on the pathogen.

Future directions and novel approaches

Our increasing knowledge of these pathways, along with emerging new technologies will allow the development of new ways to employ induced resistance in crop protection. One novel approach that is being tested is referred to as genetically engineered acquired resistance (GEAR). GEAR is a strategy for inducing SAR based on the interaction of a resistance gene and the corresponding pathogen avirulence gene (Fig. 3). When an avirulence gene is expressed in a plant containing the cognate resistance gene, the cells in which the two genes are expressed undergo HR and this in turn induces SAR. To be practical, the co-expression of the avirulence gene and the resistance gene must be limited to a small number of cells, enough to trigger small HR lesions and induce SAR, but without causing so much necrosis that it negatively affects yield. In the system being tested, three transgenes are introduced into the plant, a constitutively expressed *Avr* gene, a resistance gene driven by a constitutive promoter, but containing a non-autonomous transposable element which blocks expression, and a transposase gene expressed at low levels. The low level of transposase activity results in the excision of the non-autonomous element from the resistance gene in a small number of cells, resulting in the expression of the resistance gene. In these cells, the combination of the resistance gene product and avirulence gene product triggers HR, switching on SAR. Preliminary results are encouraging, but clearly effective use of this system will require very tight regulation of the excision event (Jones et al 1995).

The expanding field of genomics may provide valuable information that will be useful in extending our knowledge of plant defence responses. By comparing profiles of gene expression, and protein and metabolite accumulation under various induction conditions, it should be possible to gain a broader understanding of the molecular and biochemical basis of the enhanced resistance. This may provide not only more information on the known pathways, but it may

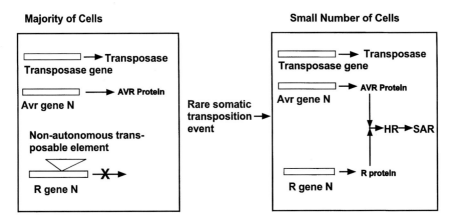

FIG. 3.   Genetically engineered acquired resistance is based on the induction of SAR following the hypersensitive cell death that occurs following the interaction between a disease resistance (R) gene and the corresponding pathogen avirulence (*Avr*) gene. By carefully controlling the expression of the resistance gene, cell death can be limited to a small number of cells. This is achieved by inserting three transgenes into the plant. The *Avr* and transposase genes are expressed constitutively, with the transposase expressed at low levels. The resistance gene contains a non-autonomous transposable element insertion that blocks expression of the R gene and the cells grow normally. In a small number of cells, the transposase causes the transposable element to excise from the R gene. The R gene can then be expressed. The R protein interacts with the AVR protein, resulting in the death of the cell. This localized cell death triggers SAR.

also lead to discovery of novel induced defence pathways. It may provide markers for induced pathways that could be used in further screens for new and improved chemical inducers. Markers can also be used in mutant screens to identify key regulatory genes controlling induction pathways. As demonstrated by the plants engineered to overexpress *NIM1/NPR1*, these regulatory genes have potential in developing lines with enhanced resistance. The *CIM* genes, which regulate the induction of multiple defence pathways, have the potential to provide an even broader spectrum of resistance than that afforded by SAR alone. The *CIM* genes may be used to enhance resistance following the example of the *NIM* transgenic plants. They may also be useful as targets for chemical screens, to identify chemicals that induce the pathways.

    In conclusion, the use of a plant's own induced defence response pathways may provide a significant contribution to total crop protection strategies. With a better understanding of the known pathways, and with the possible identification of new pathways, the development of these approaches will become more feasible. Through the novel use of components of these pathways, alone and in combination with current methods, we may be able to develop resistance strategies that are both more durable and broad spectrum.

References

Ajlan AM, Potter DA 1992 Lack of effect of tobacco mosaic virus-induced systemic acquired resistance to arthropod herbivores in tobacco. Phytopathology 82:647–651

Baldwin I 1998 Jasmonate-induced responses are costly but benefit plants under attack in native populations. Proc Natl Acad Sci USA 95:8113–8118

Cao H, Bowling SA, Gordon S, Dong X 1994 Characterization of an *Arabidopsis* mutant that is nonresponsive to inducers of systemic acquired resistance. Plant Cell 6:1583–1592

Cao H, Glazebrook J, Clarke JD, Volko S, Dong X 1997 The Arabidopsis *NPR1* gene that controls systemic acquired resistance encodes a novel protein containing ankyrin repeats. Cell 88:57–63

Cao H, Li L, Dong X 1998 Generation of broad-spectrum disease resistance by overexpression of an essential regulatory gene in systemic acquired resistance. Proc Natl Acad Sci USA 95:6531–6536

Choi D, Bostock RM, Avdiushko S, Hildebrand DF 1994 Lipid-derived signals that discriminate wound-responsive and pathogen-responsive isoprenoid pathways in plants: methyl jasmonate and the fungal elicitor arachidonic acid induce different 3-hydroxy-3-methylglutaryl-coenzyme A reductase genes and antimicrobial isoprenoids in *Solanum tuberosum* L. Proc Natl Acad Sci USA 91:2329–2333

Delaney T, Friedrich L, Ryals J 1995 *Arabidopsis* signal transduction mutant defective in chemically and biologically induced disease resistance. Proc Natl Acad Sci USA 92:6602–6606

Doares SH, Narváez-Vásquez J, Conconi A, Ryan CA 1995 Salicylic acid inhibits synthesis of proteinase inhibitors in tomato leaves induced by systemin and jasmonic acid. Plant Physiol 108:1741–1746

Doherty HM, Selevendran RR, Bowles DJ 1988 The wound response of tomato plants can be inhibited by aspirin and related hydroxybenzoic acids. Physiol Mol Plant Path 33:377–384

Feys BJF, Benedetti CE, Penfold CN, Turner JG 1994 *Arabidopsis* mutants selected for resistance to the phytotoxin coronatine are male sterile, insensitive to methyl jasmonate, and resistant to a bacterial pathogen. Plant Cell 6:751–759

Fidantsef AL, Stout MJ, Thaler JS, Duffey SS, Bostock RM 1999 Signal interactions in pathogen and insect attack: expression of lipoxygenase, proteinase inhibitor II, and pathogenesis-related protein P4 in the tomato, *Lycopersicon esculentum*. Physiol Mol Plant Pathol 54:97–114

Friedrich L, Lawton K, Ruess W et al 1996 A benzothiadiazole derivate induces systemic acquired resistance in tobacco. Plant J 10:61–70

Gaffney T, Friedrich L, Vernooij B et al 1993 Requirement of salicylic acid for the induction of systemic acquired resistance. Science 261:754–756

Glazebrook J, Rogers EE, Ausubel FM 1996 Isolation of *Arabidopsis* mutants with enhanced disease susceptibility by direct screening. Genetics 143:973–982

Gorlach J, Volrath S, Knauf-Beiter G et al 1996 Benzothiadiazole, a novel class of inducers of systemic acquired resistance activates gene expression and disease resistance in wheat. Plant Cell 8:629–643

Gundlach H, Muller MJ, Kutchan TM, Zenk MH 1992 Jasmonic acid is a signal transducer in elicitor-induced plant cell cultures. Proc Natl Acad Sci USA 89:2389–2393

Jones DA, Hammond-Kosack KE, Jones JDG 1995 Method of introducing pathogen resistance in plants. Patent No. WO 95/31564. Issued 23 Nov 1995

Karban R, Adamchak R, Schnathorst WC 1987 Induced resistance and interspecific competition between spider mites and vascular wilt fungus. Science 235:678–680

Lawton K, Uknes S, Friedrich L et al 1993 The molecular biology of systemic acquired resistance. In: Fritig B, Legrand M (eds) Mechanisms of plant defense responses. Kluwer Academic, Dordrecht, The Netherlands, p 410–420

Lawton K, Friedrich L, Hunt M et al 1996 Benzothiadiazole induces disease resistance in *Arabidopsis* by activation of the systemic acquired resistance signal transduction pathway. Plant J 10:71–82

Malamy J, Carr JP, Klessig DF, Raskin I 1990 Salicylic acid: a likely endogenous signal in the resistance response of tobacco to viral infection. Science 250:1002–1004

McConn M, Browse J 1996 The critical requirement for linolenic acid is pollen development, not photosynthesis, in an *Arabidopsis* mutant. Plant Cell 8:403–416

Metraux J-P, Signer H, Rals J et al 1990 Increase in salicylic acid at the onset of systemic acquired resistance in cucumber. Science 250:1004–1006

Metraux J-P, Ahl-Goy P, Staub T et al 1991 Induced resistance in cucumber in response to 2,6-dichloroisonicotinic acid and pathogens. In: Hennecke H, Verma DPS (eds) Advances in molecular genetics of plant–microbe interactions, vol 1. Kluwer Academic, Dordrecht, The Netherlands, p 432–439

Molina A, Hunt MD, Ryals JA 1998 Impaired fungicide activity in plants blocked disease resistance signal transduction. Plant Cell 10:1903–1914

O'Donnell PJ, Calvert C, Atzorn R, Wasternack C, Leyser HMO, Bowles DJ 1996 Ethylene as a signal mediating the wound response of tomato plants. Science 274:1914–1917

Peña-Cortés H, Albrecht T, Prat S, Weiler EW, Willmitzer L 1993 Aspirin prevents wound-induced gene expression in tomato leaves by blocking jasmonic acid biosynthesis. Planta 191:123–128

Pieterse CMJ, van Wees SCM, Hoffland E, van Pelt JA, van Loon LC 1996 Systemic resistance in *Arabidopsis* induced by biocontrol bacteria is independent of salicylic acid accumulation and pathogenesis-related gene expression. Plant Cell 8:1225–1237

Pieterse CMJ, van Wees SCM, van Pelt JA et al 1998 A novel signaling pathway controlling induced systemic resistance in *Arabidopsis*. Plant Cell 10:1571–1580

Rasmussen JB, Hammerschmidt R, Zook MN 1991 Systemic induction of salicylic acid accumulation in cucumber after inoculation with *Pseudomonas syringae* pv *syringae*. Plant Physiol 97:1342–1347

Ryals JA, Neuenschwander UH, Willits MG, Molina A, Steiner H-Y, Hunt MD 1996 Systemic acquired resistance. Plant Cell 8:1809–1819

Ryals J, Weymann K, Lawton K et al 1997 The *Arabidopsis* NIM1 protein shows homology to the mammalian transcription factor inhibitor IκB. Plant Cell 9:425–439

Ryan CA 1990 Protease inhibitors in plants: genes for improving defenses against insects and pathogens. Annu Rev Phytopathol 28:425–449

Shah J, Tsui F, Klessig DF 1997 Characterization of a salicylic acid-insensitive mutant (*sai1*) of *Arabidopsis thaliana*, identified in a selective screen utilizing the SA-inducible expression of the *tms2* gene. Mol Plant-Microbe Interact 10:69–78

Stout MJ, Workman KV, Bostock RM, Duffey SS 1998 Specificity of induced resistance in the tomato, *Lycopersicon esculentum*. Oecologia 113:74–81

Stout MJ, Fidantsef AL, Duffey SS, Bostock RM 1999 Signal interactions in pathogen and insect attack: systemic plant-mediated interactions between pathogen and herbivores of the tomato, *Lycopersicon esculentum*. Physiol Mol Plant Pathol 54:115–130

Vijayan P, Shockey J, Lévesque CA, Cook RJ, Browse J 1998 A role for jasmonate in pathogen defense of *Arabidopsis*. Proc Natl Acad Sci USA 95:7209–7214

White RF 1979 Acetylsalicylic acid (aspirin) induces resistance to tobacco mosaic virus in tobacco. Virology 99:410–412

Xu Y, Chang PL, Liu D et al 1994 Plant defense genes are synergistically induced by ethylene and methyl jasmonate. Plant Cell 6:1077–1085

DISCUSSION

Lamb: Diana Bowles has reported an antagonism between salicylate and jasmonate, but this is with respect to jasmonate function in proteinase inhibitors and the wound response. Is jasmonate function blocked by salicylate in this ISR pathway too?

Dietrich: As far as I know, that has not been done. ISR is a relatively difficult system to work with. It does not give a huge increase in resistance, and so far there are no good reporter genes for it.

Scheel: The *nim* gene doesn't need to be induced to function in the pathway. So what is the effect if NIM protein is over-expressed? Has that been studied? Do you need a certain threshold level or is it simply dose-dependence?

Dietrich: It seems to be dose-dependent. However, Xinnian Dong's group has also over-expressed *nim* and found that even though RNA levels rose 20-fold, protein levels only rose threefold at the most (Cao et al 1998). But there was definitely a correlation between protein level and level of resistance. The *nim* gene has been cloned and has weak homology to the I-κB protein, which is involved in defence signalling in mammalian systems. NIM has got some ankyrin repeats, which are protein–protein interaction domains found in I-κB. Initially, people thought that this was a parallel pathway conserved between mammals and plants. I-κB works by interacting with NF-κB: when the pathway is activated, I-κB is phosphorylated, ubiquitinated and degraded, which releases NF-κB into the nucleus, which in turn regulates transcription. With NIM, what seems to happen is that when the system is activated the NIM protein goes into the nucleus; it's not degraded at all. Regulation seems to be different, but upon activation it becomes localized in the nucleus. Therefore it looks as if it's a threshold effect: if you have a little bit more protein it's much more responsive, but you still need the salicylate or BTH. We have *nim* over-expressing plants in a *nahG* background, and they show no response to SA.

Takabayashi: Have you ever experienced gene silencing, in which the over-expressed gene suddenly becomes silent?

Dietrich: We do have some lines that show little or no enhanced resistance, and these are the lines with the lowest level of expression of the *NIM1* transgene. We have not, however, recovered any lines that are more susceptible than wild-type plants, which would be the expected result of silencing.

Takabayashi: Have you ever tried treating the plants with cytokinin? In the tobacco system, this causes over-expression of PR genes.

Dietrich: We haven't looked at that.

Gierl: Is there any concern that if you induce SAR in crops you might cause an accumulation of toxic substances in the plants?

Dietrich: Yes, that's a regulatory question that we are currently addressing. At one level, these pathways are switched on in a lot of plants anyway, so it's something that humans have been exposed to ever since they began eating plants.

Gierl: You also showed such a nice clear picture that after SAR there's no longer any infection. Are there other pathogens known that are still capable of infecting such plants?

Dietrich: There are pathogens that are not affected by SAR. There are some pathogens that seem to be regulated by the jasmonate pathway. The jasmonate mutants, for example, are much more susceptible to necrotrophic soil-borne fungi. These are non-SAR pathogens, suggesting that the jasmonate pathway is involved in defence against a different subset of pathogens.

Hammerschmidt: We have looked at the general diversity of phenolic compounds in soybean after SAR induction. We have not seen any obvious changes in secondary metabolites in soybean leaves after BTH treatment. We have also never seen any alteration in secondary metabolites in cucumber after BTH treatment. This suggests that BTH treatment does not alter secondary chemistry dramatically, resulting, for example, in the accumulation of antifungal materials. However, we do not know whether plants induced to express natural defences such as PR proteins will show any major effects on growth. This is something we are interested in.

One other observation on the potential of induced plants having repressed yields is that in the years we have been doing the soybean work we haven't seen any effect in the absence of disease where we have had yield suppression. In some cases in the absence of disease we have seen a little higher yield, which we attribute to knocking down the low grade pathogens, which don't seem to cause a problem but which reduce yields slightly. We see this to a lesser degree also in cucumber.

Schultz: If the distinction really holds up between SA and JA influence, are there characteristics in common among pathogens that are JA related and not SA related?

Dietrich: If we induce SAR biologically with pathogens we don't necessarily get this SA/JA distinction. From a biological perspective this isn't happening in nature. Salicylate is just one signal that's induced by the biological pathogen, whereas there are other signals that can induce both pathways.

Schultz: Do plant responses to pathogens that blast their way in by actually damaging cell walls resemble responses to wounding? If so, this would mean that many fungal responses would resemble wound responses, but viruses wouldn't be related to wound responses, nor would most bacteria.

Dietrich: There are some data from Rick Bostock (Fidantsef et al 1999, Stout et al 1999) indicating that there was actually a bacterial pathogen inducing both pathways.

Pickett: Has the effect on higher trophic interactions involving insects on plants that have been treated with Bion been investigated? It would be interesting to see whether there are any effects on plant volatiles and whether there is any effect on parasitoid foraging.

Boland: I can comment on that. We checked Bion for volatile induction in the lima bean system; there is no volatile induction at all. Furthermore, because Bion is an activated ester, we checked whether it is conjugated to an amino acid and transported. We made such conjugates, tested them, but they don't induce volatiles.

Dicke: We found that predatory mites don't respond to Bion-treated lima bean plants.

Firn: How long is the Bion treatment effective for? Does it have a long half-life itself in the plant, or does it induce things which have a long half-life?

Dietrich: The durability of the response is variable. In monocots such as wheat and rice you only have to treat once in the growing season. In some dicots it is not as long-lasting.

Hammerschmidt: Soybeans need four applications, cucumber one or two.

Scheel: Do you see differences in the response physiology if you compare over-expressors and genetically resistant plants?

Dietrich: I haven't done those experiments myself, but you do get a hypersensitive response.

Firn: I have a lurking unease that by inducing or augmenting a natural system of defence, you're actually increasing the selection for pathogens to overcome this defence.

Dietrich: I share your concern. One of the ideas is that because this is such a broad spectrum defence it will be difficult to overcome. There is always the potential for resistance — but this is a problem with any strategy that you take.

Pickett: A three-pronged approach with a fungicide, expression of the pathogen resistance gene and the Bion treatment would mean that there would not be so much selection pressure on each component.

Firn: But you are targeting the 'generic level' of the defensive systems which is shared by many species.

Dietrich: The response isn't exactly the same in every species. There are different patterns of gene expression. The PR genes are different in different species.

Firn: But Bion works on many different plants, so there must be a 'generic' target.

Pickett: I think Bion provides a rather enlightened approach: it is good to see industry taking such a course. However, I get the distinct impression that many companies are merely screening for signal effects from any synthetic material. This will give us just another generation of pesticides without the science having moved very far forward.

Schultz: It strikes me that there's one benefit and two risks in this approach. The benefit is that these system seem to be designed with back-ups. At this meeting we are hearing straightforward, simplified pathway descriptions, but when you read the literature you discover that actually you can get that result by this way, that way and so forth. The plant is designed that way because there is the possibility of blocking one pathway and winning the game. If you preserve that, this is a probably a more stable approach. But we're also hearing that the efficacy probably depends on maintaining the variation at the output side: the number of PR genes, variation in their expression, and that sort of thing. It is always tempting to simplify when one engineers these things. I would doubt the value or sustainability of modifying a single response, such as PR1 protein expression.

Dietrich: There are many other mutants in these pathways. The story that is emerging from double mutants and epistasis analysis is that these are definitely networks rather than linear pathways. It is not simple to work with these things.

Mitchell-Olds: Nevertheless, it may be a lot simpler and may have larger and more measurable and more useful effects if we can study and manipulate the signal transduction pathways, rather than particular downstream elements. With cold tolerance we have certainly seen that regulatory genes have large effects (Jaglo-Ottosen et al 1998), and transgenic plants for these regulatory factors have substantially greater tolerance to freezing. Whereas, if you just transgenically alter expression of one gene that does one little thing, it has disappointingly small effects from the point of view of those who would like to manipulate agricultural cold tolerances. We are beginning to understand a little bit about how the pathways work: we have no idea of how variable they are. There is some suggestion they may be somewhat conserved, and that's why people do this in *Arabidopsis* first.

Dietrich: It is true what you say about the downstream events. Initially people tried over-expressing individual PR genes and they had little or no effect. You really have to go more towards the regulatory approach.

Schultz: I wish I knew what actually stopped the pathogen in one of these systems. I would be a lot more comfortable manipulating signalling system if I knew what the hell the end point was. It drives me crazy! Yes, PR1 is a marker for resistance, but I would be a lot more comfortable if we knew what the product was while we're tinkering with the signalling.

Pickett: What do we know about PR1?

Dietrich: Its function is unknown. Some of the PR genes encode chitinases or gluconases that are probably involved in degrading hyphal cell walls.

Schultz: The operative term is 'probably' when it comes to what do they do.

Hammerschmidt: In experiments in the paper by Alexander et al (1993) where they did multiple transformations with PR genes in tobacco, they showed an incremental increase in resistance to oomycete fungi in plants transformed with PR1. Similar results were found in potato transformed with PR5 (Liu et al 1994).

Maybe this will help alleviate the fears of Richard Firn about pathogens overcoming the acquired resistance: studies on acquired resistance have shown that you not only turn on a whole series of PR genes, some of which encode products that are going to be active against certain pathogens, but you have also enhanced the potential for the expression of general defence which will certainly be multicomponent. If you transform into plants SAR defence-associated genes one at a time, you only get an incremental change in resistance. Thus it is likely that multiple mechanisms are operating together to resist infection in plants expressing SAR. The pathogen will have to overcome most or all of the defences first to overcome SAR, unless the pathogen can somehow block the initial signalling that leads to SAR expression.

Mitchell-Olds: My reading of the physiology and genetics literature is that resistance to insects and pathogens really is polygenic. The tragedy for molecular biology is that quantitative genetics is true. It's going to be really hard to manipulate.

Schultz: In the USA, transformation with Bt is promoted heavily because it does kill pests dead, after all. As an evolutionary biologist, I can understand the arguments against this, but the desire to kill pests quickly makes it difficult to overcome this movement with evolutionary reason.

Pickett: Besides the chemical ecology group in my department at Rothamsted, the main interest of the department is on resistance by insects to pesticides and their control. We are convinced that because of inappropriate deployment of Bt genes to protect crops there will be a massive resistance problem, particularly in China. Of course, by that time the Chinese cotton industry may have already caused serious competition to the American cotton industry.

Schultz: I have colleagues who are preparing to deploy Bt-engineered apples in orchards! In answer to my queries about the evolution of resistance, they say they will just use five or six different endotoxins and deploy them spatially and temporally separate, using the diversity of the end product to slow resistance evolution.

Pickett: The use of Bt genes has run away from science. First of all, there is the belief amongst some that if there is good expression all the pests will be killed, hence no resistance. This is a commonly held view in medicine and is very naïve and not relevant for agriculture. More recently, other ideas have emerged for 'pyramiding in genes' (which we don't actually have). Bt represents a very narrow range of activity that will not lead to a whole range of novel toxin genes.

Karban: What ever happened to the idea of fusing things like Bt to inducible promoters?

Pickett: Vast resources have already been expended in developing Bt genes industrially and there is consequently so much pressure to recoup profits that there is no time for sophisticated but sustainable approaches to be developed.

Schultz: It is still alive. My apple colleagues need specific tissue expression as well; they're hoping to develop inducible Bt expression in particular tissues. In apple orchards generally speaking, you don't care if the insects eat the leaves much, but you want to stop insects from scoring the apple. Thus they want expression in immature skin of the apple, but they don't want to leave it there because the apple is a food.

Pickett: There must be promoter systems that are available here because there is a range of chemistry going on only in the skin, not least of course (E,E)-α-farnesene synthesis.

Lamb: In the 'industrial' insecticide world, is there an equivalent approach to that of Novartis, looking for enhancers of the broad spectrum SAR immunity? That is, are there ways of augmenting the natural defences against insects?

Pickett: I have talked to some multinationals of the possibilities here, and they were not interested.

Lamb: One could take an entirely analogous approach here.

Pickett: I think it's a good idea.

Schultz: There seems to be interest in the jasmonate analogues.

Boland: We have analogues. The most important thing in my opinion is that we can have analogues for selective triggering of biosynthetic pathways. I don't know whether these will ever be important for pest management, but it could be important for triggering plants or cell cultures to produce useful products.

Pickett: There's no doubt about it: the key role of molecular biology in agriculture is not going to be crop protection; rather it will be quality and nutritional traits. However, if this work generates the means of switching genes associated with some of those traits, this could be very valuable.

Poppy: Colleagues of mine at Rothamsted, together with workers from INRA Versailles and Bures-sur-Yvette, have been working on an oilseed rape containing a proteinase inhibitor. The feeding behaviour of coleopteran pests on the oilseed rape was studied. The beetles were switching between cysteine and serine proteases within their guts, and thus were not being affected by the transgenic rape — in fact, they were doing better on it! When we suggested to the genetic engineers to produce a double construct, containing both the cysteine and serine protease inhibitor, they couldn't get the double construct to express reliably in the plant. The idea of pyramiding and stacking genes is all very well, but technically this is difficult.

Pickett: I'm sure the technical barriers will eventually be surmounted, but along the way we may have lost the Bt toxins through massive resistance problems. As you say, the other ones we have at present are more difficult to manage. The snowdrop lectins are not standing up at all well, for example.

Bohlmann: We speculated this morning about different signals involved in inducing volatile release after herbivory in corn and cotton, for instance. What is known about the signals involved in developmentally regulating volatile release in flowers? Could there be common elements or cross-points in the two forms of regulation?

Firn: There is a temporal regulation of flower scent releases, so it is obviously subject to regulation on the temporal level.

Poppy: There is a distinct diurnal pattern in nectar production in most species of plants that have been studied. The production of volatiles, which advertise the nectar reward, is often linked with nectar production, as is the foraging activity of bees. The peak in foraging activity on any one particular flower correlates strongly with maximum nectar production by those flowers — bees are even more plastic than parasitoids and quickly learn to exploit large nectar resources.

Turlings: For moth-pollinated plants there's also this daily rhythm of odour emissions, with the attractive volatiles only being released at specific times during the night. This contrasts with the induced volatiles that are mostly produced during the day.

Vet: There is the extra-floral nectar induction by herbivores that Felix Waeckers and the group of Joe Lewis have worked on.

Karban: A student in my lab, Anurag Agrawal, has recently published that plants that are damaged produced extrafloral nectar and become more attractive to ants, which then protect the plants (Agrawal 1998).

References

Agrawal AA 1998 Leaf damage and associated cues induce aggressive ant recruitment in a neotropical ant-plan. Ecology 79:2100–2112

Alexander D, Goodman RM, Gut-Rella M et al 1993 Increased tolerance to two oomycete pathogens in transgenic tobacco expressing pathogenesis-related protein 1a. Proc Natl Acad Sci USA 90:7327–7331

Cao H, Li L, Dong X 1998 Generation of broad-spectrum disease resistance by overexpression of an essential regulatory gene in systemic acquired resistance. Proc Natl Acad Sci USA 95:6531–6536

Fidantsef AL, Stout MJ, Thaler JS, Duffey SS, Bostock RM 1999 Signal interactions in pathogen and insect attack: expression of lipoxygenase, proteinase inhibitor II, and pathogenesis-related protein P4 in the tomato, *Lycopersicon esculentum*. Physiol Mol Plant Pathol 54:97–114

Jaglo-Ottosen KR, Gilmour SJ, Zarka DG, Schabenberger O, Thomashow MF 1998 *Arabidopsis CBF1* overexpression induces *COR* genes and enhances freezing tolerance. Science 280:104–106

Liu D, Raghothama KG, Hasegawa PM, Bressan RA 1994 Osmotin overexpression in potato delays development of disease symptoms. Proc Natl Acad Sci USA 91:1888–1892

Stout MJ, Fidantsef AL, Duffey SS, Bostock RM 1999 Signal interactions in pathogen and insect attack: systemic plant-mediated interactions between pathogens and herbivores of the tomato, *Lycopersicon esculentum*. Physiol Mol Plant Pathol 54:115–130

Future use of plant signals in agricultural and industrial crops

Richard Karban

Department of Entomology, University of California, Davis, CA 95616, USA

Abstract. Entomologists have begun to develop techniques involving vaccinations and applications of chemical elicitors to protect crops, applying what they know about the signals that mediate induced resistance. Apple and strawberry growers who tolerate some mites have realized reduced subsequent pest problems. Winegrape growers who have had chronically devastating populations of Pacific mites vaccinate their vines with less damaging Willamette mites and achieve smaller pest populations and improved yield. Vaccinations using live herbivores provide all the signals that plants use to induce resistance but are logistically difficult. Many chemicals have been identified as elicitors of induced resistance including one (benzothiadiazole) that is now marketed commercially for use against plant diseases. Exogenous application of jasmonic acid (JA) induces accumulation of many putatively defensive compounds and reduces herbivory in field situations. Ongoing experiments are evaluating the potential of exogenous JA applications in several crops. Recent evidence suggests that the pathways that plants use to protect against pathogens and against herbivores may be antagonistic, such that stimulating one depresses the other. The extent to which this trade-off may limit the ability of natural plants to defend themselves or agriculturists to limit plant parasites remains to be determined.

1999 Insect–plant interactions and induced plant defence. Wiley, Chichester (Novartis Foundation Symposium 223) p 223–238

The most widely used technique in western preventive medicine involves vaccinating susceptible patients with attenuated strains of diseases and other chemical elicitors of induced resistance. Similar techniques have been used sparingly for several decades to protect crops against plant diseases (Campbell & Macdonald 1989). These attempts have produced several notable successes although vaccination techniques have been adopted in only a small percentage of the commercial systems for which they are seemingly amenable. Recently, interest in vaccinations has increased with the identification, development and commercial deployment of chemical elicitors of induced resistance against plant pathogens. Induced resistance arises in response to damage or its cues as opposed to constitutive resistance which is always expressed. Coincident with these

223

developments by plant pathologists have been preliminary attempts to use vaccination and elicitor techniques to induce resistance in crop plants against their insect and mite herbivores. This review will consider these recent developments and speculate about future possibilities of inducing resistance against herbivores.

At this time, several different approaches appear to hold promise. The most straightforward of these involves using herbivores to activate the plant's resistance mechanisms. Using herbivores has the advantage of providing all of the actual signals that the plant has evolved to react to. However, actual herbivores are difficult to culture, apply and control. Actual damage may also cause some level of economic loss to the crop.

One alternative involves applying only the signals that the plant 'perceives'. As we learn more of the basic biology of how plants perceive herbivores and convey this information to other ramets we will be better able to induce resistance without the logistical problems associated with actual herbivores and without the damage that they cause while inducing resistance. Chemical elicitors of resistance have other advantages over actual herbivores. They can be used to manipulate the timing of induction so that it occurs prior to large increases in pest populations. They can also be used to quickly elevate plant resistance after herbivores colonize. In addition, chemical elicitors may be more effective at inducing resistance than is natural herbivory and may induce higher levels of resistance. For example, benzothiadiazole (BTH), a synthetic elicitor of the salicylic acid (SA) pathway, induced similar chemical responses as did pathogens and was found to be an extremely potent inducer (Gorlach et al 1996). Similarly, specific factors contained in insect saliva have been found to be extremely powerful inducers of plant chemicals associated with induced resistance against herbivores (Korth & Dixon 1997). These and other chemicals can be applied exogenously in controlled concentrations to crops to produce graded induced responses, some of which exceed the natural responses of the plants to pathogens and herbivores.

Classical breeding programs for plant resistance to diseases and herbivores have been limited to existing heritable variation present in the species (or closely related species). Genetic engineering offers the exciting possibility of incorporating novel resistance traits from other species, families and even phyla into crop plants with existing desirable traits. In addition, it should be possible to make a trait that is now constitutively expressed inducible by fusing it with an inducible promoter. It would then only be expressed following herbivory or following activation by a chemical elicitor. These exciting prospects are currently limited by our ability to isolate and move resistance traits. We now have the technical ability to insert genes coding for specific single products (defences) into plants that previously lacked them. This approach is likely to produce resistance that is not nearly as effective nor as durable as naturally occurring plant resistance which relies on the coordinated activation of numerous different products. For this reason, vaccinations involving live herbivores or manipulations of signals (elicitors) of resistance are more attractive techniques. However, as our technical ability to identify plant defences and transduction signals and to transform plants simultaneously with many genes increases, genetic manipulation of plants will become a more feasible and more powerful technique.

The techniques discussed above all have the advantage of relying on naturally occurring mechanisms of resistance. Deploying these defences *in planta* will therefore be as safe or as unsafe as naturally occurring resistant plants.

Since resistance traits often reduce herbivory, reduce damage to plants and increase plant fitness when herbivores are present (Karban & Baldwin 1997, Agrawal 1998, Baldwin 1998), why aren't these resistance traits expressed constitutively? The answer to this question has been elusive but it appears that resistance can be costly to plants. These costs may include resource allocation to defence, and especially life history trade-offs associated with defence and increased susceptibility to other pathogens and herbivores associated with the expression of resistance against some herbivores (Karban & Baldwin 1997, Agrawal & Karban 1998). These costs in the broad sense may favour inducible rather than constitutive expression of resistance in natural situations but costs may be partially offset by agricultural practices. For example, resistance that is costly to natural plants in terms of resources may be far less costly in agricultural situations where fertilizer and water can be supplied. In addition, induced resistance may have inherent advantages over constitutively expressed resistance (Karban et al 1997a, Agrawal & Karban 1998). These advantages can be conserved when using vaccination and elicitors to induce resistance.

Costs of resistance also represent a potential hurdle to using chemical elicitors to induce resistance to pathogens and herbivores. Several attempts to use signal molecules to induce disease resistance have been thwarted by problems of autotoxicity. In pioneering work in the mid 1980s, Kuc (1987) reported that he could elicit production of phytoalexins in unattacked green bean and soybean plants by exogenous applications of glucan elicitors derived from fungi. Unfortunately, frequent applications of elicitor were required to maintain resistance and these caused severe necrotization and stunting of growth to the crops. Similarly, isonicotinic acid (INA) was identified by scientists at Ciba as a promising elicitor of systemic resistance against many plant pathogens although it caused unacceptably high levels of phytotoxicity (Kessman et al 1994).

Vaccinations against herbivores using live herbivores

Agricultural crops support many herbivore species and only a small fraction of these cause economic losses. The activities of one herbivore species may induce resistance against a more damaging pest species; this may occur naturally or it may be deliberately encouraged by agronomic practices. There are two examples in which growers have been encouraged to ignore populations of less damaging mites and have achieved control over more damaging populations. Apple growers in Michigan, USA have controlled a serious economic pest, European red mites, by allowing less damaging apple rust mites (Croft & Hoying 1977, B.A. Croft, personal communication). Strawberry growers in New Zealand have found that by not treating two-spotted spider mites during the first growing season, mite damage in subsequent seasons was reduced (Butcher et al 1987, M. R. Butcher, personal communication). In both of these cases, induced resistance is likely to play a role, although in neither has the mechanism been elucidated.

Winegrapes in the central valley of California represent a situation where one herbivore (Willamette mites) has been deliberately encouraged in an attempt to induce resistance against a second, more damaging herbivore species (Pacific spider mites). The two species are often negatively associated; vineyards with chronic outbreaks of Pacific mites tend to lack Willamette mites and healthy vineyards tend to have low populations of Willamette mites and few or no Pacific mites (Flaherty & Huffaker 1970, English-Loeb & Karban 1988). Introductions of Willamette mites into vineyards with chronic mite problems have resulted in reduced populations of Pacific mites and have been more effective and more consistent than other control methods such as miticides and releases of predaceous mites (Fig. 1) (English-Loeb et al 1993, Karban et al 1997b). Vaccinations in problem vineyards were associated with increased yield characteristics. However, vaccinations were only successful if Willamette mites were present early in the growing season as new shoots expanded (Hougen-Eitzman & Karban 1995). Willamette mites persisted in untreated vineyards so that a single introduction was sufficient to produce sustainable control over several seasons (Karban et al 1997b). The negative effects of Willamette mites on populations of Pacific mites were as strong when the two species were separated spatially or temporally as when they co-occurred, with or without predators, suggesting induced resistance as a mechanism (Hougen-Eitzman & Karban 1995).

Although this vaccination technique has provided inexpensive and consistently effective results in commercial vineyards, it has not become widely adopted. There are no commercial suppliers of Willamette mites. Many growers cannot distinguish between the two mite species. Most importantly, the technique lacks an effective facilitator — someone who can make a profit by encouraging it. At the very least, many growers now regard small populations of herbivorous Willamette mites as beneficial and foregoing miticides for their control may allow Willamette mites to replace Pacific mites in these vineyards. Evidence of varying quality suggests that select herbivores in many other crops can cause induced resistance and vaccinations

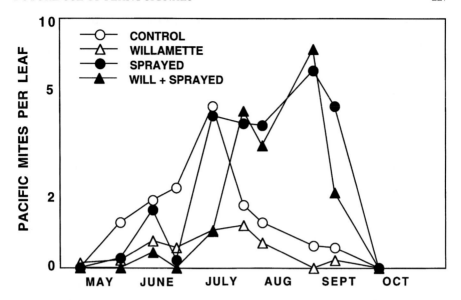

FIG. 1.   Densities of economically damaging Pacific mites in a vineyard managed with four different strategies. Only vaccinations of Willamette mites maintained low populations of Pacific mites throughout the season. Applications of miticide sprays (dicofol) alone or with Willamette mites were ineffective and resulted in high populations through much of the season. (Data from Karban et al 1997b.)

probably could be effective in some of these (see Table 6.1 in Karban & Baldwin 1997).

Protection against herbivores by the use of chemical elicitors

Many chemicals have been found to induce resistance against pathogens and herbivores (Lyon et al 1995, Karban & Baldwin 1997). These elicitors produce no direct antibiotic effects but work by inducing resistance in the target plants. These chemicals are easier to produce, distribute, store and apply compared with live vaccinations. Plant signals are highly conserved so that adapting these as elicitors potentially could allow them to be used in many diverse crop and pest situations (Constabel & Ryan 1998, Morris et al 1998). However, to date field tests of elicitors of induced resistance have been attempted only a few times.

Several elicitors of induced resistance against pathogens have been tested and released for agricultural use (Kendra et al 1989, Ye et al 1995). The most exciting of these endeavors is the commercial production of BTH. This material has no direct antimicrobial activity although it activates the same biochemical and molecular markers as seen when actual pathogens are used as inducers (Gorlach et al 1996). It has been marketed initially by Novartis in Europe under the tradename 'Bion' to protect wheat against powdery mildew, although it confers broad spectrum resistance against diverse pathogens of crops ranging from tobacco to maize (Morris et al 1998). This large-scale commercial experiment will provide a sense of whether elicitors of induced resistance are likely to become the 'pesticides of the future'.

Chemical elicitors of induced resistance against herbivores have received much less attention. Jasmonic acid (JA) and its volatile methyl ester, methyl jasmonate, have been identified as natural signal molecules in the responses of plants to herbivory. They are found in many plants and they are synthesized in response to wounding. According to current models, herbivory releases systemin, which moves throughout the plant and initiates a cascade of reactions including formation of JA and ultimately the activation of genes encoding compounds that provide defence (Schaller & Ryan 1996). Exogenous treatment of plants with these powerful jasmonate elicitors induced the accumulation of chemicals involved in plant defence (Staswick 1992, Reinbothe et al 1994, Karban & Baldwin 1997). Mutant plants that could not mount the jasmonate response did not accumulate putatively defensive compounds following damage and experienced low fitness in laboratory environments with herbivores (Howe et al 1996, McConn et al 1997).

Exogenous application of minute quantities of jasmonates simulated the damage caused by herbivores and increased resistance to herbivores in the laboratory (Constabel et al 1995, Thaler et al 1996). This was likely the work of induced resistance as jasmonates incorporated into artificial diets caused no direct toxicity to herbivores (Avdiushko et al 1997). However, techniques that provide successful control of herbivores in the laboratory often fail in the field.

Recent results indicate that endogenously applied JA induced resistance against all the common herbivores of tomato plants under conditions of production agriculture over three field seasons in Davis, California (J. S. Thaler, M. J. Stout, R. Karban & S. S. Duffey, unpublished data). Densities of noctuid caterpillars, flea beetles, aphids and thrips were all reduced on plants treated with levels of JA that produced no phytotoxicity. Induced resistance killed early instars of the caterpillars, adult flea beetles avoided induced plants, and mechanisms responsible for the effects on aphids and thrips remain unknown. Treated plants suffered 60% less leaf damage than did control plants (Thaler 1999). However, because tomato plants grown in Davis are very tolerant of foliar herbivory, these large effects did not translate into measurable differences in yield (Thaler 1999).

Other plants are far less tolerant of herbivory. We are currently experimenting with applications of JA to control herbivores on a variety of other agricultural crops. Preliminary results suggest possible control against herbivores on lettuce and carrots (C. A. Black, R. Karban & L. D. Godfrey, unpublished data) and against phylloxera on grapevines (A. Omer, J. S. Thaler, J. Granett & R. Karban, unpublished data).

Interactions between signal pathways

The simplest model of signal pathways posits that SA induces resistance against pathogens and JA induces resistance against herbivores. If this model was accurate, then we could activate either or both of these two response systems by applying chemical elicitors of these pathways under the appropriate environmental conditions (threats from pathogens or herbivores). Unfortunately, the actual situation is far more complex. For one thing, the signal pathways appear to be dependent upon environmental conditions. For another, these two signal pathways interact with each other in complex ways, and possibly with other plant hormones such as ethylene and abscisic acid used to signal abiotic stresses.

In some instances, exogenous treatment with JA has induced resistance to pathogens (Schweizer et al 1997) and treatment with BTH occasionally induces resistance to insects (Inbar et al 1998). Evidence is accumulating that the two pathways are often antagonistic. Applications of SA or its mimics depress the ability of plants to mount induced responses to herbivores and applications of JA depress induced responses to pathogens.

SA has been known for some time to inhibit defensive responses to herbivores. Treatment with SA prevented wounded plants from accumulating putatively defensive proteinase inhibitors (Doherty et al 1988). This occurred because SA blocked the signal pathway that culminates in synthesis of JA and transcription of genes coding for plant defences against herbivores (Doares et al 1995). To determine whether applications of BTH, a mimic of SA, would depress induced responses to herbivores, we treated field-grown tomato plants with: (a) JA alone ($3\,\mu$M per plant); (b) BTH alone ($2.5\,\mu$M per plant); (c) JA and BTH; and (d) neither JA nor BTH (J. S. Thaler, M. J. Stout, R. Karban & S. S. Duffey, unpublished data). Plants of all four treatments were challenged with *Spodoptera exigua* caterpillars. Survival of caterpillars was reduced on plants treated with JA relative to controls. Survival was enhanced on plants treated with BTH and those on plants treated with both JA and BTH showed intermediate rates of survival. Applications of BTH interfered with the plant's ability to protect itself from herbivores and lessened the potential benefits of applying elicitors to stimulate induced resistance to herbivores.

Recent evidence also indicates that applications of JA depress resistance to pathogens. We subjected field-grown tomato plants to the four elicitor treatments described above and challenged with bacterial speck agent, *Pseudomonas syringae* pv. tomato (J. S. Thaler, A. L. Fidanstef, S. S. Duffey & R. M. Bostock, unpublished data). Plants sprayed with BTH had the fewest bacterial speck lesions and JA-treated and control plants had the most lesions. Those sprayed with both JA and BTH had an intermediate number of lesions. These results indicate that applications of JA depress natural or chemically elicited systemic acquired resistance (SAR) in tomato plants.

The trade-offs between the two signalling pathways observed in the field bioassays described above were also observed at the level of gene expression. Plants that were treated with BTH had increased mRNA levels for a pathogenesis-related (PR) protein (P4), an indicator of resistance to pathogens, and reduced mRNA levels for a proteinase inhibitor (PINII), an indicator of resistance to herbivores (Fig. 2; Fidantsef et al 1999). Conversely, plants that were treated with JA expressed reduced mRNA levels for the PR protein but increased mRNA levels for the proteinase inhibitor. Recent work suggests possible mechanisms for the connections between the two pathways. Tobacco

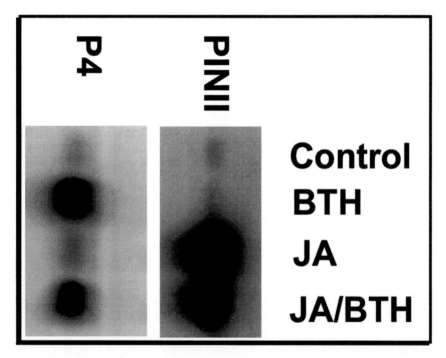

FIG. 2.   Induction of putative defensive chemicals of induced resistance against herbivores (PINII) and pathogens (P4). Treatment of tomato leaves with BTH induced high mRNA levels for the pathogenesis-related protein P4 but no induction of the proteinase inhibitor PINII. Treatment with JA induced high mRNA levels for PINII but not P4. Treatment with both BTH and JA resulted in attenuation of gene expression relative to levels when the elicitors were applied alone. (Data from Fidantsef et al 1999.)

plants that were transformed with a GTP-encoding gene increased their levels of SA (rather than JA) in response to wounding (Sano & Ohashi 1995). According to the model developed by this group, a GTP-binding protein in the cell membrane acts as a molecular switch that directs the wound signal to either the SA or JA pathway (Sano & Ohashi 1995, Seo et al 1997). Levels of cytokinins may also be involved in regulating or switching the flow of these signals. In any case, with rather little modification, one stimulus is able to cross over to activate either cascade.

Conclusion

Plants in nature or agriculture face a diversity of challenges that includes both pathogens and herbivores. The antagonism that we have observed between these two signalling pathways may limit the ability of plants to defend themselves simultaneously against both types of parasites. Certainly, the use of vaccinations and elicitors of induced resistance may offer exciting and novel means of controlling the defences that a crop plant expresses. However, the antagonism between the two signalling pathways may represent a firm limit to the level of control that we can exert. The antagonism may also represent an ecological cost associated with either of these two defensive systems (*sensu* Karban & Baldwin 1997); plants that are defended against some parasites may be necessarily less well defended against others. For this reason, inducible defences may be favoured over constitutive ones. Inducibility allows the plant to determine which of the two responses is most appropriate to the pathogen or herbivore regime that it encounters, thereby minimizing the ecological costs of expressing resistance to parasites that are not encountered. The same benefit should also be available to growers who can select the elicitors that will modify the plant's phenotype to match closely with its environment.

References

Agrawal AA 1998 Induced responses to herbivory and increased plant performance. Science 279:1201–1202

Agrawal AA, Karban R 1998 Why induced defenses may be favored over constitutive strategies in plants. In: Tollrian R, Harvell CD (eds) Ecology and evolution of inducible defenses. Princeton University Press, Princeton, NJ, p 45–61

Avdiushko SA, Brown GC, Dahlman DL, Hildebrand DF 1997 Methyl jasmonate exposure induced insect resistance in cabbage and tobacco. Environ Entomol 26:642–654

Baldwin IT 1998 Jasmonate-induced responses are costly but benefit plants under attack in native populations. Proc Natl Acad Sci USA 95:8113–8118

Butcher MR, Penman DR, Scott RR 1987 Population dynamics of two spotted spider mites in multiple year strawberry crops in Canterbury. NZ J Zool 14:509–517

Campbell R, Macdonald RM 1989 Microbial inoculation of crop plants. Oxford University Press, Oxford (Special Publication of the Society for General Microbiology)

Constabel CP, Ryan CA 1998 A survey of wound-induced and methyl jasmonate-induced leaf polyphenol oxidase in crop plants. Phytochemistry 47:507–511

Constabel CP, Bergey DR, Ryan CA 1995 Systemin activates synthesis of wound-inducible tomato leaf polyphenol oxidase via the octadecanoid defense signaling pathway. Proc Natl Acad Sci USA 92:407–411

Croft BA, Hoying SA 1977 Competitive displacement of *Panonychus ulmi* (Acarina: Eriophyidae) by *Aculus schlechtendali* (Acarina: Eriophyidae) in apple orchards. Can Entomol 109: 1025–1034

Doares S, Syrovets T, Weiler E, Ryan CA 1995 Oligogalacturonides and chitosan activate plant defense genes through the octadecanoid pathway. Proc Natl Acad Sci USA 92:4095–4098

Doherty HM, Selvendran RR, Bowles DJ 1988 The wound response of tomato plants can be inhibited by aspirin and related hydroxybenzoic acids. Physiol Mol Plant Pathol 33:377–384

English-Loeb GM, Karban R 1988 Negative interactions between Willamette mites and Pacific mites: possible management strategies for grapes. Entomol Exp Appl 48:269–274

English-Loeb GM, Karban R, Hougen-Eitzman D 1993 Direct and indirect competition between spider mites feeding on grapes. Ecol Appl 3:699–707

Fidantsef AL, Stout MJ, Thaler JS, Duffey SS, Bostock RM 1999 Signal interactions in pathogen and insect attack: expression of lipoxygenase, proteinase inhibitor II, and pathogenesis-related protein P4 in the tomato, *Lycopersicon esculentum*. Physiol Mol Plant Pathol 54:115–130

Flaherty DL, Huffaker CB 1970 Biological control of Pacific mites and Willamette mites in San Joaquin valley vineyards. I. The role of *Metaseilus occidentalis*. Hilgardia 40:267–308

Gorlach J, Volrath S, Knauf-Beiter G et al 1996 Benzothiadiazole, a novel class of inducers of systemic acquired resistance, activates gene expression and disease resistance in wheat. Plant Cell 8:629–643

Hougen-Eitzman D, Karban R 1995 Mechanisms of interspecific competition that result in successful control of Pacific mites following inoculations of Willamette mites on grapevines. Oecologia 103:157–161

Howe GA, Lightner J, Browse J, Ryan CA 1996 An octadecanoid pathway mutant (JL5) of tomato is compromised in signaling for defense against insect attack. Plant Cell 8:2067–2077

Inbar M, Doostdar H, Sonoda RM, Leibee GL, Mayer RT 1998 Elicitors of plant defensive systems reduce insect densities and disease incidence. J Chem Ecol 24:135–149

Karban R, Baldwin IT 1997 Induced responses to herbivory. University of Chicago Press, Chicago, IL

Karban R, Agrawal AA, Mangel M 1997a The benefits of induced defenses against herbivores. Ecology 78:1351–1355

Karban R, English-Loeb G, Hougen-Eitzman D 1997b Mite vaccinations for sustainable management of spider mites in vineyards. Ecol Appl 7:183–193

Kendra DF, Christian D, Hadwiger LA 1989 Chitosan oligomers from *Fusarium solani*/pea interactions, chitosan/β-glucanase digestion of sporelings and from fungal wall chitin actively inhibit fungal growth and enhance disease resistance. Physiol Mol Plant Pathol 35:215–230

Kessmann H, Staub T, Hofmann C et al 1994 Induction of systemic acquired disease resistance in plants by chemicals. Annu Rev Phytopathol 32:439–459

Korth KL, Dixon RA 1997 Evidence for chewing insect-specific molecular events distinct from a general wound response in leaves. Plant Physiol 115:1299–1305

Kuc J 1987 Plant immunization and its applicability for disease control. In: Chet I (ed) Innovative approaches to plant disease control. Wiley, New York, p 255–274

Lyon GD, Reglinski T, Newton AC 1995 Novel disease control compounds: the potential to 'immunize' plants against infection. Plant Pathol 44:407–427

McConn M, Creelman RA, Bell E, Mullet JE, Browse J 1997 Jasmonate is essential for insect defense in *Arabidopsis*. Proc Natl Acad Sci USA 94:5473–5477

Morris SW, Vernooij B, Titatarn S et al 1998 Induced resistance responses in maize. Mol Plant-Microbe Interact 11:643–658

Reinbothe S, Mollenhauer B, Reinbothe C 1994 JIPs and RIPs: the regulation of plant gene expression by jasmonates in response to environmental cues and pathogens. Plant Cell 6:1197–1209

Sano H, Ohashi Y 1995 Involvement of small GTP-binding proteins in defense signal-transduction pathways of higher plants. Proc Natl Acad Sci USA 92:4138–4144

Schaller A, Ryan CA 1996 Systemin — a polypeptide defense signal in plants. BioEssays 18: 27–33

Schweizer P, Buchala A, Silverman P, Seskar M, Raskin I, Metraux J 1997 Jasmonate-inducible genes are activated in rice by pathogen attack without a concomitant increase in endogenous jasmonic acid levels. Plant Physiol 114:79–88

Seo S, Sano H, Ohashi Y 1997 Jasmonic acid in wound signal transduction pathways. Physiol Plant 101:740–745

Staswick PE 1992 Jasmonate, genes, and fragrant signals. Plant Physiol 99:804–807

Thaler JS 1999 Jasmonic acid mediated induced plant resistance in agriculture: effects on herbivory and yield in tomato plants. Environ Entomol 28:30–37

Thaler JS, Stout MJ, Karban R, Duffey SS 1996 Exogenous jasmonates simulate insect wounding in tomato plants (*Lycopersicon esculentum*) in the laboratory and field. J Chem Ecol 22:1767–1781

Ye XS, Strobel N, Kuc J 1995 Induced systemic resistance (ISR): activation of natural defense mechanisms for plant disease control as part of integrated pest management. In: Reuveni R (ed) Novel approaches to integrated pest management. CRC Press, Boca Raton, FL, p 95–113

DISCUSSION

Pickett: I didn't understand the treatment in which you combined the Willamette mite with the miticide. What was the point behind this?

Karban: We wanted to determine whether our technique was compatible with other practices. It was a disaster. The miticide was very disruptive and killed the Willamette and predatory mites. In almost every instance it caused the problem to be much worse. However, the growers have really gotten on the treadmill with these miticide treatments.

Pickett: So did you do this treatment to demonstrate that use of the miticide would take out the advantageous population of Willamette mites?

Karban: That gives us a lot more credit than we are due. We did it just to compare our technique with the technique the growers were using. This was a particularly dramatic example of the disastrous effects of the miticide.

Boland: You showed data where you applied both the Willamette mite and later the miticide. Does this mean that the resistance was not induced before you killed all the mites? How long does the induction of resistance take?

Karban: We haven't done many experiments that manipulate the timing, because this system is one that we have worked with mostly in the field, and so each experiment takes a whole season. However, we know that timing is critical: for resistance to be induced effectively, the Willamette mites must be present as the shoots are elongating. If you wait a week or two, you will have missed the window (Hougen-Eitzman & Karban 1995). On the other hand, the Willamette mites are native in this system and if you introduce them into a vineyard, and don't do other stupid things such as applying miticide, they're sustainable and should provide control over many seasons (Karban et al 1997).

Dicke: You said that no one is interested in rearing Willamette mites to introduce them. This surprises me, because in biological control as soon as money can be made from rearing animals, people do it. I guess here it's more the psychological factor that growers just don't want to use what looks like a pest-in-first method.

Karban: The problem is that although many of the predatory mites can be reared commercially, Willamette mites are specialists on grapes and are not amenable to rearing in the lab.

Haukioja: I guess there is an inherent problem that once the mites are established they don't need to be bought any longer. This makes rearing them a fairly unprofitable business.

You showed that when the Willamette mites had been introduced the sugar content of grapes went up. Why is this?

Karban: The conventional wisdom is that Pacific mites slow down photosynthetic rates, so the vines are not producing as much carbohydrate. The level of sugar in the grapes is the largest economic concern. Presumably, reducing the numbers of Pacific mites alleviates this problem. I don't know how good the evidence for this mechanism is.

Pickett: Are organic farmers interested in using natural JA treatment to get these beneficial effects?

Karban: I think so. They're looking for new techniques like most of the growers are.

Pickett: Would that have to be under licence of Bud Ryan's patent (Ryan et al 1991), or does spraying it on as you have done avoid this infringement?

Karban: I don't know. Our JA sprays are all on experimental crops so that we don't need to worry about the patent laws. Perhaps one way of getting round the patenting issues would be for a company to develop some of the analogues that Wilhelm Boland was describing yesterday.

Pickett: If it was going to be sold in competition with pesticides, there may need to be a bigger effect. But if it went into the organic market, these farmers would probably be quite happy with the effect you have demonstrated. The product would also have to be natural.

Schultz: We're getting a great deal of interest in this technique from growers in Pennsylvania.

Pickett: Is that from conventional or organic growers?

Schultz: It is actually both.

Is the failure to approve of biotic induction to protect crops because of the visible costs of induction, or is it because 60% foliage loss doesn't matter to tomatoes?

Karban: Maybe some of both. Tomatoes are highly tolerant of leaf damage. What we are looking at is a net cost: this includes whatever benefits there might be from not being eaten by herbivores along with whatever cost there might be in terms of reduced growth. We don't see those kinds of costs when we look for reductions in growth.

Schultz: We don't either. My graduate student has done exactly the same experiment, and in greenhouse tomatoes protected from herbivores, there is little cost of inducing proteinase inhibitors or alkaloids. There are subtle shifts in terms of number of seeds in fruit and things like that.

Pickett: Do the tomatoes taste better?

Schultz: Not better, but they definitely taste good — my laboratory has more tomatoes than we know what to do with! Since we count the seeds, we have to smash all these tomatoes and do something with them, so we have the most wonderful salsa supply you have ever seen.

Pickett: We have to destroy all food crops used in such research.

Schultz: We don't: we benefit from the control treatments of many of our agricultural experiments.

Pickett: We have some amazing laws in Europe these days.

Schultz: I'm quite surprised that there is no production cost, because JA is a growth inhibitor.

Dicke: Has anything been published on these lack of effects on growth? I noticed a tendency to be reluctant even to think about using JA because of its potential negative physiological effects.

Karban: Yes, this paper is in press in *Environmental entomology* (Thaler 1999).

Takabayashi: You sprayed the JA and found that the number of all herbivore species decreased. What is the mechanism behind this? Is it because of low performance of the herbivores on the JA-treated plants, or the increased activity of their natural enemies on these plants?

Karban: For two of the herbivores — flea beetles and spodoptera caterpillars — we know a little bit of the answer. In the case of flea beetles, they're choosy and prefer not to land on plants that have been induced. For the spodoptera caterpillars, survivorship is reduced. For the other two herbivores, aphids and the thrips, we don't know.

Baldwin: And are those alterations in survivorship due to parasitism rates or are they direct effects?

Karban: Direct effects, by and large.

Boland: Are there phytoalexins which build up after this JA treatment?

Karban: We have not looked.

Boland: What was the lowest concentration of JA you applied in the field?

Karban: I believe it's 0.5 mM, sprayed so that the tops of the plants looked wet.

Schultz: We get significant results from 0.1 mM to 10 mM, over the whole range, but there's the potential for damage to the plant if we get much above 1 mM.

Scheel: Was the combined treatment with BTH and jasmonate simultaneous, or did you try sequential treatments? This might make a big difference.

Karban: At this point, the only experiments that we have done have been with simultaneous applications. One of the things that is on the schedule for this coming field season is to try different combinations and sequences of these materials.

Dietrich: Was the jasmonate treatment a single treatment?

Karban: Yes.

Dietrich: What did you score for the insects?

Karban: We gave a single treatment early in the season, and then monitored the numbers of insects throughout the season.

Schultz: We've done this experiment spraying weekly over the entire growth of the plant, and under those conditions we still don't get any negative impact on the plant growth.

Vet: You monitored the herbivores: did you also monitor their natural enemies on these plants? Do you see effects on the presence or performance of natural enemies?

Karban: We have some interesting results in the tomato system that we are just now coming to grips with. Jennifer Thaler has found that herbivores on induced plants have higher rates of parasitism.

Schultz: We're in the same spot and we're getting similar patterns.

Städler: Did you see any negative results? For example, did you ever see that there are more pests?

Karban: I'm sure we're going to turn those sorts of things up as we keep looking. In other crops in which we have done just a quick screening we certainly see negative results. But in the tomato system, these are all the herbivores that we've got, and JA has been effective against all of them.

Lamb: You have shown that salicylate might interfere with jasmonate synthesis. Work from Yuko Ohashi's lab in which they have titrated exogenous salicylate and jasmonate, has shown the induction of different sets of PR genes and mutual antagonism of the effects of jasmonate and salicylate (Niki et al 1998). This implies, unless there is some sophisticated feedback, that there is a site downstream and salicylate interferes with jasmonate action as well as synthesis. What is known about where jasmonate hits salicylate synthesis and action?

Karban: We don't know anything about how it might work here. We have just seen the phenomenon.

Haukioja: Is it known whether JA has any secondary effects on plants, such as making them more or less susceptible to environmental stress?

Karban: JA does a lot of things: the list of these kinds of secondary effects is enormous. We summarized the varied effects of JA in our book (Karban & Baldwin 1997, p 96–100).

Firn: One JA treatment seems to produce a long-lasting effect. Does this mean that JA has a long environmental half-life?

Karban: I don't know.

Lamb: BTH is not the active principle, but is metabolized to give controlled release of the active principle.

Dietrich: BTH is metabolized when it gets into the plant, to the free acid form and other metabolites. Both the applied form and probably some of the metabolites are active in inducing SAR. But I'm not sure how stable the active forms are *in planta.*

Pickett: The follow-on from that would be to try the methyl thioester of JA, which might give better cuticular penetration and then release JA within the plant, as did the S-methyl ester of the benzomethiodiazole.

Boland: The penetration of a surface is important. We also see this with the indanone derivatives. We usually apply these as methyl esters or allyl esters. Both are more or less activated esters which can be hydrolysed within the cell. We do know that the free acid is essential for provoking any response, but the free acid is less active if you apply it as such. Penetration of a surface by a neutral component is an important aspect.

Pickett: This is known as the ion trap effect, because the phloem system has a higher pH. Thus a weak acid, when it is associated, or an ester, are sufficiently lipophilic to enter the plant, but when the acid is ionized in the phloem it is held there, giving both phloem loading and translocatability.

Dicke: I would like to come back to the question on the mechanism, because I'm surprised that if you have one application early in the season there is protection throughout the season. When you apply JA, do you find proteinase inhibitors at the end of the season in the new foliage that has been produced after the JA application? These compounds are known to be made systemically when JA is applied, but what happens to the newly produced tissues?

Karban: The chemical response dissipates over the season, but the biological effects seem to persist. The biological and chemical effects last for at least six weeks. After this time it is hard to detect chemical differences between the control and induced plants, but by that time the tomato plants are spatially very complicated.

Schultz: In that case, I would imagine that the major life history components, including growth, feeding and reproduction, are over for the big herbivores, and the major portion of the damage is averted.

Karban: Except for damage to the fruit. Many of these herbivores have many generations over a single growing season.

Schultz: A single application definitely raises the critical chemical traits to a level that will reduce performance of the 4–8 week herbivore life history.

Lamb: In SAR, classical physiological work showed that one immunization shot would get the plant right through the life cycle, i.e. the induced immunity was reasonably durable.

References

Hougen-Eitzman D, Karban R 1995 Mechanisms of interspecific competition that result in successful control of Pacific mites following inoculations of Willamette mites on grapevines. Oecologia 103:157–161

Karban R, Baldwin IT 1997 Induced responses to herbivory. University of Chicago Press, Chicago, IL

Karban R, English-Loeb G, Hougen-Eitzman D 1997 Mite vaccinations for sustainable management of spider mites in vineyards. Ecol Appl 7:183–193

Niki T, Mitsuhara I, Seo S, Ohtsubo N, Ohashi Y 1998 Antagonistic effect of salicylic acid and jasmonic acid on the expression of pathogenesis-related (PR) protein genes in wounded mature tobacco leaves. Plant Cell Physiol 39:500–507

Ryan CA, Farmer EE, inventors; Washington State University Research Foundation, asignee 1991 12 Dec. Method of inducing plant defence mechanisms. International Patent Application Number PCT/US91/03865

Thaler JS 1999 Jasmonic acid mediated induced plant resistance in agriculture: effects on herbivory and yield in tomato plants. Environ Entomol 28:30–37

Genetics and evolution of insect resistance in *Arabidopsis*

Thomas Mitchell-Olds

Max-Planck-Institut für Chemishe Ökologie, Tatzendpromenade 1a, D-07745 Jena, Germany

Abstract. The genetic and molecular tools available in *Arabidopsis* allow identification of insect resistance genes. Many functional aspects of pest recognition and signal transduction are conserved in the defensive physiology of a broad range of plant species. Therefore, studies of insect resistance in *Arabidopsis* may be extended to functional genomics studies in many plant species of agricultural and ecological importance. Because of public concerns for field release of genetically modified organisms, naturally occurring genetic variation for resistance to insect herbivores will be valuable in plant breeding. Combined studies employing QTL mapping and candidate resistance genes are necessary to find and understand the genes responsible for variation in resistance. We review experiments showing that plant populations contain high levels of genetic variation for defensive physiology and disease and insect resistance, and that this variation can be manipulated to alter resistance and its components in a predictable fashion. In *Arabidopsis*, we can map the genes controlling physiological variation, and estimate the importance of regulatory or enzyme-encoding loci. Finally, we review functional genomics approaches for identification of insect resistance genes in *Arabidopsis*.

1999 Insect–plant interactions and induced plant defence. Wiley, Chichester (Novartis Foundation Symposium 223) p 239–252

Arabidopsis thaliana provides an efficient experimental system to study physiological mechanisms and molecular genetics of insect resistance in plants (Bevan et al 1998, Meyerowitz & Somerville 1994, Settles & Byrne 1998). In addition to the usual tools of plant physiology and biochemistry, *Arabidopsis* provides genetic methods to clone any gene that causes measurable phenotypic variation. *Arabidopsis* has a dense linkage map of molecular markers, and physical maps of large-insert clones (bacterial and yeast artificial chromosomes: BACs and YACs) are nearly complete for all five *Arabidopsis* chromosomes. DNA sequence from expressed sequenced tags (ESTs), which provide partial sequence data from anonymous cDNAs, is available from approximately 15 000 expressed genes. Since *Arabidopsis* has a total of approximately 21 000 genes, it is therefore estimated that DNA sequence information is available for approximately three-quarters of all genes. Genome sequencing is also progressing rapidly: one third of the genome has been sequenced, and the complete 100 megabases of non-repetitive DNA should be completed in less than three years. The chromosomal locations of many genes are known from linkage mapping studies or from genomic sequencing. Chromosome walking allows positional cloning of mutant genes. Transformation is efficient and straightforward, so over-expression and knockouts (caused by antisense, co-suppression, or insertional mutagenesis) can be used to study gene function. Large scale insertional mutagenesis procedures using transposons and T-DNAs have been employed by many laboratories. Finally, international stock centres provide genetic and molecular resources to the *Arabidopsis* community, including mutants, ecotypes collected from the field, DNA clones, etc.

It is important to consider whether insect resistance mechanisms found in *Arabidopsis* can be extended to plant species of ecological or agricultural interest. Clearly, the generality of these results will depend upon the species being considered, and may vary among different signal transduction pathways or defensive metabolites. Since little is known about the molecular basis of insect resistance, we will also consider molecular biology of resistance to microbial pathogens, which is much better understood.

When considering the generality of resistance mechanisms, physiological plant defence against insects and microbial pathogens can be divided into three conceptual phases: recognition of attack, signal transduction and deployment of defences.

Insect, nematode, and pathogen recognition genes have many similarities among diverse plant species. Proteins in the leucine-rich repeat (LRR) gene family have been implicated in resistance to *Pseudomonas* in *Arabidopsis*, *Puccinia* resistance in *Hordeum*, viral resistance in *Nicotiana*, and aphid and nematode resistance in tomato (Hammond-Kosack & Jones 1997, Milligan et al 1998, Rossi et al 1998). Gene duplication and natural selection have resulted in a wide diversity of recognition genes (Parniske et al 1997, Wang et al 1998). Although these resistance genes can evolve rapidly, their common heritage is recognizable from conserved LRR domains. Unfortunately, relatively little is understood about the recognition phase of plant defence against insect herbivores, so few generalizations can be made at this time.

Signal transduction pathways functioning in plant defence are beginning to be understood in *Arabidopsis* and other plant species. Clearly, some of these signal transduction pathways are conserved among many plant species. For example, jasmonic acid (JA) signalling has been described in conifers, grasses, *Arabidopsis* and many other dicotyledonous plants. However, the downstream genes and metabolites induced via this pathway differ among plant species. For example, methyl jasmonate induces nicotine in *Nicotiana*, glucosinolates in *Brassica*, taxol in *Taxus*, and O-methyltransferase in *Hordeum* (Baldwin 1998, Du et al 1995, Lee et al 1997, Mirjalili & Linden 1996). However, our ability to examine conservation or diversity of signal transduction pathways among different species is biased by experimental factors. Using tools of modern molecular biology, it is straightforward to document the existence of a previously described signalling pathway in a new species. However, it is much more difficult to characterize a novel signalling pathway in plant species which lack the genetic tools of model organisms such as *Arabidopsis*. Therefore, it is easier to document conserved pathways in different species, than to discover novel pathways that lack a broad taxonomic distribution.

Finally, in the phase of deployment of defence mechanisms, we find examples of conserved defences, as well as substantial variation among plant groups. For example, proteinase inhibitors are widespread in monocot and dicot species (Koiwa et al 1997). On the other hand, a vast array of novel secondary compounds are found in a variety of plant species (Rosenthal & Berenbaum 1992).

In summary, our ability to extrapolate information on the molecular basis of insect resistance from *Arabidopsis* to related plant species will vary among pathways and target species. Presumably most resistance mechanisms from *Arabidopsis* will function similarly in closely related *Brassica* species (Price et al 1994), including oilseed rape and cruciferous vegetables. Clearly, some signal transduction pathways will have similar function in more distantly related species. For example, because JA signalling is conserved across a broad range of plant species (Creelman & Mullet 1997), we anticipate that *Arabidopsis* signal transduction genes can be used to clone and study homologues in cereal crops. Because of the many genetic and molecular tools available for *Arabidopsis*, and the availability of a complete genomic sequence in a few years, *Arabidopsis* provides an efficient system for understanding plant defence against insect enemies in natural populations and in agriculture.

Natural variation

Knowledge regarding the molecular basis of insect resistance in plants will be useful for understanding ecology and evolution in natural populations, and for manipulating insect resistance in agricultural plant species. In both situations, genes that are polymorphic within or among closely related species are of primary interest. For example, evolutionary ecologists have documented substantial genetic variation for resistance within natural plant populations. Since susceptibility to insect herbivory is deleterious in many environments (Marquis 1992), why is resistance genetically polymorphic within populations? Costs of resistance provide one possible explanation (Simms 1992). *Ecological costs* occur if genetic resistance to one pest species also causes genetic susceptibility to a different natural enemy. Alternatively, *allocation costs* occur if genetic investments in defence are energetically expensive, thus reducing individual fitness in the absence of insect herbivores. In both instances, natural genetic variation in levels of resistance is central to cost arguments. The central question in plant–pest interactions is not whether costs exist, but whether costs of resistance are responsible for evolutionary variation in susceptibility to natural enemies.

Naturally occurring genetic variation for resistance to insect herbivores may also be valuable in plant breeding and agriculture. Public perception of field trials with genetically modified organisms (GMOs) is unsympathetic in many countries (e.g. Schiermeier 1998). Agricultural production of GMOs requires extensive involvement with public relations, government regulation and oversight, and physical security surrounding field trials. Consequently, naturally occurring genetic variation may provide a useful and publicly acceptable alternative to some methods in genetic engineering.

For these reasons, ecologists, evolutionary biologists and plant breeders seek to understand the genes responsible for variation in insect resistance, and the physiological and historical factors that explain extant genetic variation. Because of this interest in the causes and consequences of naturally occurring variation, the past few years have seen a remarkable convergence of methods and questions among human medical genetics, *Drosophila* population genetics, and studies of natural and agricultural plant populations. Each of these disciplines has found extensive genetic variation for physiological and developmental traits, which have been related to polymorphisms at known genes (Hanson et al 1996, Long et al 1998, Mitchell-Olds & Pedersen 1998, Nickerson et al 1998). Expanding genomics information in each of these organisms provides similar tools for genetic analysis: facile linkage mapping, candidate genes that may control traits of interest, and increasing quantities of genomic sequence which places linked markers and candidate genes within a known chromosomal context. Because of scientific and political concerns for release of GMOs, especially those carrying resistance genes, plant breeders must now face the same questions that concern medical geneticists and pharmaceutical companies: which naturally occurring genetic polymorphisms influence health, disease and other quantitative traits?

In well characterized, genetically tractable plant species, two approaches contribute to finding the genes that control variation in insect resistance. We can combine linkage mapping of quantitative trait loci (QTL) with functional understanding of candidate resistance genes in chromosomal regions of interest (Mitchell-Olds & Pedersen 1998). Alone, neither method is sufficient to identify the actual loci responsible for insect resistance. Together, however, these two methods are sufficient to identify genes controlling resistance variation in *Arabidopsis* and related species.

First, QTL mapping can identify chromosomal regions that contain putative resistance genes, and provide some idea of the number, magnitude, pleiotropy and regulation of genes controlling insect resistance. The advantage of QTL mapping is that it can identify the chromosomal location of completely unknown genes. However, initial QTL mapping data are too crude to identify the actual locus corresponding to a QTL, because chromosome regions identified by QTL mapping usually contain hundreds of genes (e.g. Bevan et al 1998).

The second approach contributing to identification of loci responsible for resistance is analysis of candidate genes (e.g. Pigliucci 1996). Often we have a priori information on molecular or physiological function of known genes that contribute to defensive physiology or resistance traits. Association studies (Long et al 1998) examine whether allelic variation at these candidate genes contributes to quantitative variation, by examining association of alleles at candidate genes with resistance or its physiological components. Candidate gene studies have the advantage of examining a known locus that may actually cause insect resistance.

However, the candidate gene approach also has some disadvantages. Our knowledge of development and physiology of insect resistance is very incomplete, so only a limited subset of potentially important loci can be examined. Fortunately, differential display, expression profiling, EST sequencing and physiological studies can provide extensive information on candidate genes that may influence resistance to insect herbivores (Benedetti et al 1998, Desprez et al 1998, Mithen 1992, Newman et al 1994, Titarenko et al 1997). QTL complementation tests (Mackay & Fry 1996) may infer whether QTLs correspond to particular candidate loci, but such tests require mutant alleles for each candidate locus examined.

To avoid the limitations inherent in QTL mapping and candidate gene approaches, unified studies employing both methods are essential. First, QTL mapping can identify chromosomal regions of demonstrated importance for insect resistance. In some ways, the main advantage of QTL mapping is to eliminate most of the genome from subsequent consideration and to narrow the list of candidate genes to a tractable number of loci in particular chromosomal regions. These regions can be examined in more detail using candidate genes, fine-scale mapping, complementation tests, transformation, and statistical tests for neutral DNA sequence variation within and between species.

Genetic variation for physiology and resistance in *Brassica*

To study the functional genetic relationship between plant secondary chemistry and insect resistance, we collected *Brassica rapa* genotypes from a natural population in Montana (Siemens & Mitchell-Olds 1998). Two generations of artificial selection were conducted to increase or decrease concentrations of glucosinolates, secondary metabolites that are known to be biologically active against insect species. Artificial selection experiments were also replicated in two separate years. High- and low-glucosinolate genotypes were randomized together, and resistance to flea beetle feeding was measured on plants grown in the field. In both years, statistically significant differences in resistance to flea beetle feeding were observed between the high and low glucosinolate genotypes (1994: $F = 6.00$, $P = 0.043$; 1995: 11.17, $P = 0.002$; $n > 500$ each year). Flea beetles caused greater damage on high-glucosinolate genotypes in both experiments. These findings are in agreement with physiological studies of flea beetles and some other insects, where glucosinolates have been shown to stimulate feeding.

In a second experiment, we studied genetic variation in quantitative levels of enzyme activity for chitinase, an enzyme that has been shown to be functionally important for plant defence against fungal pathogens. This time we conducted artificial selection on plant resistance to a fungal pathogen, *Leptosphaeria maculans*. This replicated selection experiment to alter resistance also caused chitinase activity to increase by 13% in resistant genotypes. This correlated response to selection occurred even though selection was directed only at resistance, but not at chitinase activity (Mitchell-Olds et al 1996). These results indicate a functional role for naturally occurring quantitative genetic variation in resistance to fungal disease in natural plant populations.

In both experiments we observed predictable changes in physiological pathways in accord with a priori information on defensive physiology. Selection on physiology caused evolutionary changes in insect resistance in the wild, and selection on disease resistance caused genetic changes in underlying physiological defence mechanisms. These studies demonstrate the predictive and explanatory power of a functional approach to plant–herbivore and plant–pathogen interactions. We have documented high levels of genetic variation for physiological traits in plant populations, and fitness consequences of this physiological variation (Mitchell-Olds et al 1996).

Mapping QTLs controlling physiological variation

Genetic variation in physiology provides an experimental system where functional information on biochemistry, development, and gene regulation can contribute to evolutionary understanding. Interpretation of physiological genetic variation may be simpler than for other quantitative traits. Genes controlling plant physiological variation belong to two conceptual categories: 'structural genes' such as enzyme-encoding loci and associated proteins, and 'regulatory genes', a diverse category of genes that influence expression of enzyme encoding loci (Holton & Cornish 1995).

Regulatory factors may exert simultaneous coordinated control on several physiological traits; such pleiotropic effects cause a *genetic correlation* between activity levels of several enzymes (Clark & Keith 1988). Alternatively, when physiological variation is attributable to allelic differences at enzyme-encoding loci, it may be due to coding changes that alter the amino acid sequence of enzymes, or to *cis*-acting regulatory variation at the enzyme locus. In physiological genetics, QTLs that affect several related enzymes in a pathway suggest a regulatory factor that jointly controls expression of these enzymes.

We used a recombinant inbred mapping population of *Arabidopsis thaliana* to examine the genetic basis of physiological variation in primary and secondary metabolism, conducting more than 11 000 quantitative measurements of enzyme activity (Mitchell-Olds & Pedersen 1998). At 56 cM on chromosome 4 we mapped QTL influencing activity levels of GPD, PGI and G6P, suggesting that enzymes with closely related metabolic functions are co-regulated. We are now fine-mapping the QTL influencing this physiological variation. Fig. 1 shows a three megabase region containing these QTL on chromosome 4. Two megabases of this region is already sequenced, and contains 390 genes. We screened a recombinant inbred population of 300 lines, and obtained recombinants which separate this region into blocks containing, on average, about 25 genes. In this way, fine-scale mapping can move very close to the genes controlling physiological variation, as well as resistance to insect herbivores.

In the Department of Genetics at the Max-Planck-Institute for Chemical Ecology we are using the molecular and genetic tools of *Arabidopsis* to examine the functional basis of plant resistance to insect herbivores. Four approaches are being used to identify genes responsible for insect resistance. (1) Analysis of gene expression in response to insect herbivory by differential display, EST sequencing from subtracted cDNA libraries, and expression profiling with DNA microarrays.

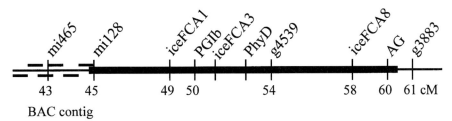

2 Megabase Sequenced Region

FIG. 1.   Fine mapping of quantitative trait loci (QTL) controlling quantitative activity levels of three glycolytic enzymes. Below the line: position on chromosome 4 in cM. Above the line: molecular markers used in early fine-scale mapping. The heavy line indicates the 2 mb sequenced region and an unsequenced BAC contig is shown to the left.

Subsequently, we will use functional genomics approaches (over-expression and knockouts) to examine gene function. (2) To isolate mutants with altered insect resistance, we have screened 11 000 EMS-mutagenized plants for altered herbivory by diamondback moth (*Plutella xylostella*). These putative mutants are being verified by several generations of progeny testing. Genes influencing insect resistance will be isolated by positional cloning. (3) We have shown that there is substantial genetic variation among ecotypes for resistance against two herbivores, diamondback moth and flea beetles. QTL mapping of insect resistance genes is in progress. (4) We are using positional cloning of candidate genes influencing biologically active secondary compounds (glucosinolates). Together, these four approaches will allow us to identify the genes that influence insect resistance in *Arabidopsis* and related species.

Acknowledgements

This work was supported by grant DEB-9527725 from the US National Science Foundation and by the Max-Planck Gesellschaft. Thanks to Swetlana Dix for secretarial assistance.

References

Baldwin IT 1998 Jasmonate-induced responses are costly but benefit plants under attack in native populations. Proc Natl Acad Sci USA 95:8113–8118

Benedetti C, Costa C, Turcinelli S, Arruda P 1998 Differential expression of a novel gene in response to coronatine, methyl jasmonate, and wounding in the *Coi1* mutant of *Arabidopsis*. Plant Physiol 116:1037–1042

Bevan M, Bancroft I, Bent E et al 1998 Analysis of 1.9 Mb of contiguous sequence from chromosome 4 of *Arabidopsis thaliana*. Nature 391:485–488

Clark AG, Keith LE 1988 Variation among extracted lines of *Drosophila melanogaster* in triacylglycerol and carbohydrate storage. Genetics 119:595–607

Creelman R, Mullet J 1997 Biosynthesis and action of jasmonates in plants. Annu Rev Plant Physiol Plant Mol Biol 48:355–381

Desprez T, Amselem J, Caboche M, Hofte H 1998 Differential gene expression in *Arabidopsis* monitored using cDNA arrays. Plant J 14:643–652

Du L, Lykkesfeldt J, Olsen C, Halkier B 1995 Involvement of cytochrome P_{450} in oxime production in glucosinolate biosynthesis as demonstrated by an *in vitro* microsomal enzyme system isolated from jasmonic acid-induced seedlings of *Sinapis alba* L. Proc Natl Acad Sci USA 92:12505–12509

Hammond-Kosack K, Jones J 1997 Plant disease resistance genes. Annu Rev Plant Physiol Plant Mol Biol 48:575–607

Hanson MA, Gaut BS, Stec AO et al 1996 Evolution of anthocyanin biosynthesis in maize kernels: the role of regulatory and enzymatic loci. Genetics 143:1395–1407

Holton TA, Cornish EC 1995 Genetics and biochemistry of anthocyanin biosynthesis. Plant Cell 7:1071–1083

Koiwa H, Bressan RA, Hasegawa PM 1997 Regulation of protease inhibitors and plant defense. Trends Plant Sci 2:379–384

Lee J, Vogt T, Hause B, Lobler M 1997 Methyl jasmonate induces an O-methyltransferase in barley. Plant Cell Physiol 38:851–862

Long A, Lyman R, Langley C, Mackay T 1998 Two sites in the delta gene region contribute to naturally occurring variation in bristle number in *Drosophila melanogaster*. Genetics 149: 999–1017

Mackay TFC, Fry JD 1996 Polygenic mutation in *Drosophila melanogaster*: genetic interactions between selection lines and candidate quantitative trait loci. Genetics 144:671–688

Marquis RJ 1992 The selective impact of herbivores. In: Fritz RS, Simms EL (eds) Plant resistance to herbivores and pathogens: ecology, evolution, and genetics. University of Chicago Press, Chicago, IL, p 301–325

Meyerowitz EM, Somerville CR 1994 *Arabidopsis*. Cold Spring Harbor Laboratory Press, Cold Spring Harbor, NY

Milligan S, Bodeau J, Yaghoobi J et al 1998 The root knot nematode resistance gene *Mi* from tomato is a member of the leucine zipper, nucleotide binding, leucine-rich repeat family of plant genes. Plant Cell 10:1307–1320

Mirjalili N, Linden JC 1996 Methyl jasmonate induced production of taxol in suspension cultures of *Taxus cuspidata*: ethylene interaction and induction models. Biotechnol Prog 12:110–118

Mitchell-Olds T, Pedersen D 1998 The molecular basis of quantitative genetic variation in central and secondary metabolism in *Arabidopsis*. Genetics 149:739–747

Mitchell-Olds T, Siemens D, Pedersen D 1996 Physiology and costs of resistance to herbivory and disease in *Brassica*. Entomol Exp Appl 80:231–237

Mithen R 1992 Leaf glucosinolate profiles and their relationships to pest and disease resistance in oilseed rape. Euphytica 63:71–83

Newman T, de Bruijn F, Green P et al 1994 Genes galore: a summary of methods for accessing results from large-scale partial sequencing of anonymous *Arabidopsis* cDNA clones. Plant Physiol 106:1241–1255

Nickerson D, Taylor S, Weiss K et al 1998 DNA sequence diversity in a 9.7 kb region of the human lipoprotein lipase gene. Nat Genet 19:233–240

Parniske M, Hammond-Kosack K, Golstein C et al 1997 Novel disease resistance specificities result from sequence exchange between tandemly repeated genes at the Cf-4/9 locus of tomato. Cell 91:821–832

Pigliucci M 1996 How organisms respond to environmental changes: from phenotypes to molecules (and vice versa). Trends Ecol Evol 11:168–173

Price RA, Al-Shehbaz IA, Palmer JD 1994 Systematic relationships of *Arabidopsis*: a molecular and morphological approach. In: Meyerowitz E, Somerville C (eds) *Arabidopsis*. Cold Spring Harbor Laboratory Press, Cold Spring Harbor, NY, p 7–19

Rosenthal GA, Berenbaum MR 1992 Herbivores: their interactions with plant metabolites, vol 1. Academic Press, San Diego, CA

Rossi M, Goggin F, Milligan S, Kaloshian I, Ullman D, Williamson VM 1998 The nematode resistance gene *Mi* of tomato confers resistance against the potato aphid. Proc Natl Acad Sci USA 95:9750–9754

Schiermeier Q 1998 German transgenic crop trials face attack. Nature 394:819

Settles AM, Byrne M 1998 Opportunities and challenges grow from *Arabidopsis* genome sequencing. Genome Res 8:83–85

Siemens D, Mitchell-Olds T 1998 Evolution of pest induced defenses in *Brassica* plants: tests of theory. Ecology 79:632–646

Simms EL 1992 Costs of plant resistance to herbivory. In: Fritz RS, Simms EL (eds) Plant resistance to herbivores and pathogens: ecology, evolution, and genetics. University of Chicago Press, Chicago, IL, p 392–425

Titarenko E, Rojo E, Léon J, Sánchez-Serrano J 1997 Jasmonic acid-dependent and acid-independent signaling pathways control wound-induced gene activation in *Arabidopsis thaliana*. Plant Physiol 115:817–826

Wang G, Ruan D, Song W et al 1998 *Xa21D* encodes a receptor-like molecule with a leucine-rich repeat domain that determines race-specific recognition and is subject to adaptive evolution. Plant Cell 10:765–779

DISCUSSION

Poppy: I assume that the flea beetles you use are field collected, and therefore will show tremendous genetic variability. In contrast, *Plutella* is very much a lab animal in which strains can easily be created. *Plutella* can therefore be adapted to a range of conditions: in Rothamsted we have susceptible strains and resistant strains to insecticides and so on. The *Plutella* you are working with will be much more genetically fixed than the flea beetles which are going to be all over the place.

Pickett: Is your flea beetle a specialist on crucifers (Brassicaceae)?

Mitchell-Olds: It's the one that we find primarily in Montana, *Phyllotreta cruciferae*, which is a specialist crucifer feeder.

Pickett: That explains why there is a positive correlation with glucosinolate levels.

We would expect it to be like our flea beetle, *Psylliodes chrysocephala*, in that it is selective for the catabolites for certain glucosinolates: principally the 3-butenyl and the 4-pentenyl. Down-regulating genes for their biosynthesis would be something we would like to do with you or Richard Mithen's group at the John Innes Institute. One of our discoveries that might be useful to you is that methyl jasmonate causes induction of one part of the glucosinolate pathway — i.e. the indolyl glucosinolates (Doughty et al 1995).

Mitchell-Olds: Those data are coming from oilseed rape. It is not clear that we are going to find the same thing in *Arabidopsis*.

Pickett: These plants may not be the same in this respect, but it might allow you to pick out which genes are which, certainly if you want to be selective between the different glucosinolates. Genetic engineering of the whole glucosinolate pathway could be difficult. Removal would be disadvantageous, of course. The first examples of double nought (00) oilseed rape cultivars with low glucosinolates were destroyed by pigeons and slugs and any other unadapted herbivore or pathogens.

Lamb: In your analysis of the glycolytic enzyme variation, the data you showed were for three enzymes that catalyse near-equilibrium reactions *in vivo*. Why would the plant care if those levels are up 10-fold even? The question is more what happens to phosphofructokinase or something like that?

Mitchell-Olds: There is a whole area in physiological genetics that deals with this sort of question. Metabolic control theory predicts there could be large changes in physiology which cause only minor changes in overall flux or fitness. However, in the real world this equilibrium theory often may not apply. For example, we know that glycolysis is up- and down-regulated by environmental factors. We are working with these steps in glycolysis primarily because we want to address a different question: is quantitative trait variation attributable to regulatory genes or to enzyme-encoding loci? Therefore, the fact that we get these three QTLs right in one place is suggestive of a regulatory locus. So far, however, we really don't know what we're seeing here.

One of our conclusions is that it's really hard to learn biology from QTL mapping. When you get down to a 5 or 10 cM region you haven't learned that much. In that figure I showed, we expect there to be about 600 genes, so there is ample room for multiple *trans*-acting factors, each affecting a single trait at a time, or perhaps a pleiotropic regulator. We just don't know yet, and that's why we are pursuing that region.

On chromosome 5 we have a PGM activity QTL. In that case, we know the flux control co-efficient, so we can predict the effect of the QTL on enzyme activity and on flux. We predict about a 3% effect on flux in that instance.

Lamb: These measurements are presumably V_{max} per unit of tissue. By measuring that you are bound to pick up regulatory loci rather than structural loci.

Mitchell-Olds: This PGM QTL appears to be right on top of the PGM-encoding locus. So far, our fine-scale mapping data are consistent with the hypothesis that this QTL may be attributable to the PGM-encoding locus.

Lamb: It could be a *cis*-acting element in the promoter.

Mitchell-Olds: Yes, we have also found another glycolytic QTL influencing hexokinase activity, where the QTL is attributable to regulatory variation in the promoter or another tightly linked regulatory site.

Obviously, one must streamline one's experimental methods in order to obtain 11 000 quantitative assays of enzyme activity. Our published experimental methods contain several caveats that might result from these streamlined assays. The amazing thing is that, in spite of all this, we can make successful predictions. We can select on physiology, and resistance changes as a correlated response to selection, which is a pleasant surprise.

Lamb: But I think that would be a really nice experiment to do. I don't know which genes you have looked at in secondary metabolism. You could take phenylalanine ammonia-lyase and 3-hydroxymethylglutaryl-Co A, which are known to be rate determining steps. Your QTL would then relate back to some specific metabolic regulatory mechanism. The phenylpropanoid pathway is a particularly strong platform on which to do that experiment.

Haukioja: Because *Arabidopsis* is an annual plant, seed set is a good measure of fitness. How much is known about correlates between plant fitness and the existence of those genes and their expression?

Mitchell-Olds: In related areas of secondary metabolism, we did an experiment where we selected up and down on total peroxidase activity in *Brassica*. We replicated these selection experiments in three separate populations, and in every case we found high peroxidase activity is associated with significantly reduced growth rate.

Gierl: Have you tested whether the resistance in your 25 resistant lines is monogenic?

Mitchell-Olds: No, we are in the third generation of this progeny testing, so we haven't started the crossing yet.

Takabayashi: You selected plants that were resistant and susceptible to the diamondback moth. Did you count the number of seeds produced by those plants to see whether they have different fitness from the wild-type or not?

Mitchell-Olds: We haven't done this in *Arabidopsis*, but we have data on that in *Brassica*, and the overall fitness effects seem to differ from year to year.

Pickett: Where exactly do the genes studied by Richard Mithen's group (i.e. that give chain elongation amino acids) fit in with your work? These are involved in the first committed step of the glucosinolate pathway.

Mitchell-Olds: In my paper I don't talk at all about glucosinolates in *Arabidopsis*. However, we do know that the glucosinolate elongase locus is on chromosome 5. That is the gene we believe we have cloned.

Scheel: In your differential display, did any gene show up that was only responsive to insects?

Mitchell-Olds: Not so far.

In the differential display the plants are all of the same age, but at the time of harvest they have either never seen an insect or have seen insects for 10 hours, or for thirty hours. Basically, it is only control versus differential periods with insects.

Scheel: So you didn't always include jasmonate level?

Mitchell-Olds: Not in the differential display, but once we have those genes we intend to check jasmonate and salicylate induction. I showed you a few Northern blots showing jasmonate effects for several genes.

One other advantage I didn't mention is the availability of mutants. A recent paper in *Plant Cell* (Pieterse et al 1998) showed a beautiful example of one of the additional advantages of doing this sort of research in *Arabidopsis*. They worked out the interplay between the jasmonate signalling pathway and the salicylate signalling pathway. But what experiments were required to get this paper in *Plant Cell*? They had a biotic challenge to the plant, and they used four mutants that they got in the mail. You write the introduction and conclusions, send it to *Plant Cell*, and start the next experiment. Availability of these pre-existing resources is a big advantage.

Baldwin: I have a question about your mutant selection lines. One of the nice things about doing mutant selection lines with pathogens is that it is easy to make sure that the billions of spores that you sprayed on the *Arabidopsis* have an equal probability of attacking the plant. You get a lot of statistical power with such experiments. It is a little hard to do this with *Plutella*. How rapidly were you able to select resistant lines with *Plutella* as compared to what people have done with pathogens? How do these selection differences compare?

Mitchell-Olds: The selection was in *Brassica*, whereas documentation of genetic variation was done in *Arabidopsis*. We haven't done any artificial selection in *Arabidopsis*. The reason why pathogens are so far along and insects are not, is because insects don't grow on petri plates and they don't give you a yes/no answer. They are harder to work with. The way round this is the way that experimental biologists have always done these things in ecology and agriculture: full randomization, controls and so on. We put known numbers of insects on randomized plots of plants and we replicated heavily.

Lamb: Related to that, one of the things in pathology is a genetic framework: the gene-for-gene interaction that governs host resistance. Is there anything like that with insects?

Mitchell-Olds: From the literature, there's the example of Hessian fly. If there are additional examples, Jack Schultz will know them, I'm sure.

Schultz: They're rare, presumably because selection on the participants is more diffuse — more things influence the interaction than the genetics of the two participants. The Hessian fly and rice gall midge do have such relationships with their host plants. This suggests to me that galling insects may be good models for this. The ability to elicit a gall in a plant would appear to involve the insect's ability to manipulate the gene expression in the plant. I would expect to see some of those gene-for-gene relationships, if it is going to exist in insect–plant interactions, in galling insects.

Lamb: In setting up the *Arabidopsis* pathosystems, most of the useful genes regarding recognition were identified from natural variation among ecotypes rather than by a mutational approach. My thought would be that for the insect resistance, the mutational approach might be better.

Mitchell-Olds: Strictly from a point of view tractability, which approach is going to give us genes with effects that are large enough to study? I don't know yet. There are clearly huge differences among the ecotypes. But whether individual QTLs will be large enough to clone is still unknown. Our mutational screen isn't quite to the point of mapping and cloning mutants yet.

With regard to major gene effects that we see in plant–pathogen interactions, quite a few papers contain caveats that genes of small effect were difficult to study, and therefore were not pursued further. There is an experimental bias to study genes of large effect, for perfectly good reasons of tractability.

Dicke: I have a comment on the difference between pathogens and herbivores in measuring resistance. With herbivores you can measure many different aspects. In an applied project we screened for resistance against spider mites among gerbera cultivars. Plant breeders are interested in checking oviposition rates of spider mites. We measured the whole life history of the spider mites and it turned out that there were small differences in oviposition rates while there were large differences in developmental rates. These differences had a large effect on intrinsic rate of population increase (r_m). Looking only at oviposition rates implies concentrating on the wrong trait.

Mitchell-Olds: Speaking as a recovering quantitative geneticist, it's useful to simplify the plant genetics, so you can say there's a defined single-gene difference, and ask about its effect on feeding or oviposition of different insects. Plant genetics and insect biology are so complicated that you have got to simplify at least one of them.

Karban: Marcel Dicke, I'm very interested in your comment. Did you find that there was a correlation between how resistant the plants looked when you just measured oviposition compared to when you measured all the other traits?

Dicke: Effects on the oviposition rate showed the same trends as effects on developmental time, but with oviposition rate we did not find as many significant differences (Krips et al 1999).

Schultz: Addressing the overall philosophical point that Tom would like to make about the use of a model system like this, I think there are wonderful benefits to be had. In particular, I really like having a genome to plunder for genes to exploit in other plants. There are some things that are difficult to achieve with that system, and we used tobacco and tomato and even cucumber as well. However, with the aspects of life history traits and metabolism in woody plants, for example, there's nothing to plunder in *Arabidopsis*. Thus there will be inevitable limits to what we can achieve there, but it is certainly a wise place to start.

Mitchell-Olds: The wild relatives of *Arabidopsis* have an amazing diversity. For example, a timber company is interested in *Arabis lignifera* because they want to study the molecular biology of lignification, which might provide useful information for genetic manipulation of forest trees.

References

Doughty KJ, Kiddle GA, Pye BJ, Wallsgrove RM, Pickett JA 1995 Selective induction of glucosinolates in oilseed rape leaves by methyl jasmonate. Phytochemistry 38:347–350

Krips OE, Witul A, Willems PEL, Dicke M 1999 Intrinsic rate of population increase of the spider mite *Tetranychus urticae* on the ornamental crop gerbera: intraspecific variation in host plant and herbivore. Entomol Exp Appl, in press

Pieterse C, van Wees S, van Pelt J et al 1998 A novel signaling pathway controlling induced systemic resistance in *Arabidopsis*. Plant Cell 10:1571–1580

Exploiting insect responses in identifying plant signals

John A. Pickett, Keith Chamberlain, Guy M. Poppy* and Christine M. Woodcock

*Biological and Ecological Chemistry Department, *Entomology and Nematology Department, IACR-Rothamsted, Harpenden, Hertfordshire, AL5 2JQ, UK*

Abstract. Plants produce compounds which can act as semiochemicals, i.e. as signals modifying the behaviour and development of other organisms without having direct physiological activity, and which generally have extremely benign toxicological properties. Such semiochemicals include plant stress signals and possibly also phytopheromones associated with the induction of defence systems but which may also be exploited by colonizing organisms. Since it is now possible to use sophisticated electrophysiological techniques to investigate insect interactions with phyto-pheromones, the detection and characterization of such signalling systems is greatly facilitated, particularly when combined with behavioural techniques for confirming the biological relevance of the signals. As well as the contribution of these studies to the understanding of plant–insect and plant–plant interactions, plant signals are potentially valuable in the regulation of gene expression for improved or alternative approaches to crop protection. It would also be possible to use these signals to 'switch on', when required, other genes associated with beneficial agronomic or nutritional crop traits.

1999 Insect–plant interactions and induced plant defence. Wiley, Chichester (Novartis Foundation Symposium 223) p 253–265

There is a demand for new methods of pest control due to the real or imagined problems with current pesticides, the rising world population and the need to provide substantially more food than can be produced with current technologies and land space. Ever increasing efforts are being directed at studying insect–plant interactions, with the aim of identifying mechanisms by which insects colonize plants, in the hope that novel control measures can be developed. Behaviourally active volatile emissions from plants play a major role in the initial colonization stages and are part of a group of bioactives termed 'semiochemicals' (from the Greek root σημειον = semeion), meaning 'sign-chemical'. Many ubiquitous plant components contribute to host attraction and, for some taxa, specific compounds have been identified as key compounds in directing insects towards their host plants (Visser 1986). Some semiochemicals may act between plants, as phytopheromones. External plant–plant signals recently proposed are the volatile methyl esters of jasmonic (Farmer & Ryan 1990, Doughty et al 1995, Baldwin 1998) and salicylic acids (Pettersson et al 1994, Shulaev et al 1997), and nitric oxide (Delledonne et al 1998).

Non-host interactions

For aphids (Aphididae), detection of host semiochemicals is mainly via olfactory organs, the 'primary rhinaria', on the fifth and sixth antennal segments. It is possible to record signals from the whole antenna (the electroantennogram, or EAG), or to insert finely sharpened tungsten electrodes into the rhinaria to record from individual neurons (single cell recording, or SCR). By linking these electrophysiological preparations directly to high resolution quartz capillary gas chromatography (GC), active components in plant extracts can be pinpointed for subsequent identification by GC-mass spectrometry (GC-MS) and comparison with authentic compounds. Many of the neurons in the primary rhinaria respond to ubiquitous plant components, often highly specifically, and some respond to compounds typical of certain plant taxa. However, some olfactory neurons are found which do not respond to either. Therefore, volatile materials from plants that are not suitable hosts for the insects in question were examined and indeed, some such cells were found to be extremely sensitive to chemicals from non-host plants. For the black bean aphid, *Aphis fabae*, it could readily be seen that there are olfactory neurons in the primary rhinarium on the fifth antennal segment which have high specificity for organic isothiocyanates, typical of non-hosts in the Brassicaceae (= Cruciferae). Interestingly, when these were compared with the same type of cell in the cabbage aphid, *Brevicoryne brassicae*, similar dose–response curves were obtained for a range of organic isothiocyanates, the 4-pentenyl homologue being the most active for both species. However, *B. brassicae* colonizes brassicaceous plants, which release isothiocyanates, whereas *A. fabae* does not. These isothiocyanates were subsequently found to repel *A. fabae* and also to inhibit the normal attractancy of host plant volatiles (Nottingham et al 1991). Investigation of other non-host plants for *A. fabae*, including members of the Lamiaceae (= Labiatae), identified a number of electrophysiologically active compounds including methyl salicylate and (1R,5S)-myrtenal, both of which were found to inhibit the attractancy of bean volatiles (Hardie et al 1994). (1R,5S)-Myrtenal is formally an oxidation product of (1S,5S)-α-pinene (Scheme 1). Although the significance of this relationship has not yet been established, it may be that aphids such as *A. fabae* can avoid a range of inappropriate plants and ecosystems by responding negatively to a compound structurally, and probably biosynthetically, closely related to monoterpene hydrocarbons such as the

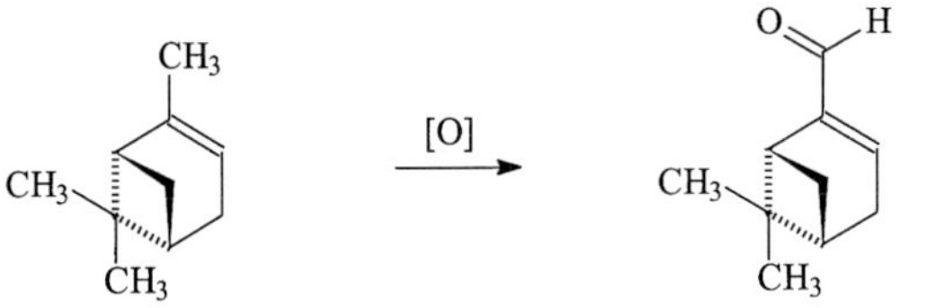

Scheme 1.   Oxidation of pinene to myrtenal.

pinenes, which are typical of many non-host plants, including the labiates and the wide range of gymnosperms.

Complementary studies on cereal aphids, including the grain aphid, *Sitobion avenae*, and the bird-cherry-oat aphid, *Rhopalosiphum padi*, the major disease vector for cereals (Poaceae = Gramineae), showed that they also possess olfactory cells specific for methyl salicylate. This compound was able to inhibit the normal attractancy of cereal volatiles in the olfactometer. In addition, it has been possible to use slow release formulations of methyl salicylate to reduce significantly aphid colonization of cereal crops, particularly barley, in the field (Pettersson et al 1994, Pickett et al 1997). Methyl salicylate is, of course, the methyl ester of salicylic acid (Scheme 2). This has a number of signalling roles within plants and is produced as a consequence of part of the phenylalanine ammonialyase (PAL) pathway, which is inducible and is responsible for many forms of chemical and structural defence (Ward et al 1991). Thus, it may be that insects responding negatively to methyl salicylate may be doing so as a result of detecting defences associated with the PAL pathway, the methyl salicylate being a volatile external form of the internal salicylate signal (Pettersson et al 1994, Shulaev et al 1997). Indeed, around 40 different species of insects, from four different Orders, have now been found to have highly sensitive electrophysiological responses to methyl salicylate. Where analogues have been investigated, for example ethyl salicylate, at least an order of magnitude in activity is lost. It is therefore likely that methyl salicylate is a general signal for insects in determining the appropriateness of plants as hosts, particularly where induced defence could make colonization uneconomic in terms of energy expended against the induced defence, and therefore to be avoided.

Scheme 2.   Methylation of salicylic acid.

The search for other plant signals

When it was shown that methyl salicylate is used as a signal by insects, it was suggested that plants could themselves respond to this compound (Pettersson et al 1994). Recently, Raskin's group (Shulaev et al 1997) has shown that methyl salicylate can indeed act as an external plant signal inducing defence against pathogens. Since it is difficult to identify chemicals that might influence plants and yet, for insects, the activity of methyl salicylate was readily demonstrated by exploiting the high sensitivity of the electrophysiological preparation, it was proposed that other plant signals might be detected by using this system. These studies are ongoing, but clear success for this strategy has already been observed. One example involves the blackcurrant-lettuce aphid, *Nasonovia ribis-nigri*. This aphid feeds on lettuce in the summer, when reproduction is asexual, or parthenogenetic, but in order to go through the sexual stage, to mate and lay eggs, it must return to the primary or winter woody host, which is blackcurrant, *Ribes nigrum* (Saxifragaceae). In the case of *R. padi*, it was already known that volatiles from the primary host, although attractive in the autumn to the sexual forms, were repellent to the spring migrants, ensuring that they moved back to the secondary or summer host. Indeed, methyl salicylate plays a role in this behaviour (Pettersson et al 1994). Volatiles of *R. nigrum* were therefore collected by air entrainment from intact plants and analysed by coupled GC-EAG, using the summer form of *N. ribis-nigri*. A range of GC peaks were associated with high EAG activity; a number of these were ubiquitous plant compounds, as expected, and some were associated specifically with *R. nigrum*. One compound which, although present at an extremely low level, showed very high EAG activity, is also known to be released during plant stress. Subsequently, this compound was shown in olfactometer tests to be repellent to *N. ribis-nigri* and also to the damson-hop aphid, *Phorodon humuli*. Behavioural tests on predators and parasitoids of aphids showed that the compound was also an attractant for the seven-spot ladybird, *Coccinella septempunctata* (Coccinellidae), and *Aphidius ervi* (Braconidae), a parasitoid of the pea aphid, *Acyrthosiphon pisum*. Excitingly, when plants were enclosed for 24 h in air containing low levels of this compound and then tested 48 h later in a wind tunnel, significantly more parasitoids oriented towards the treated plant (32%) compared with the untreated control (10%) in a dual choice bioassay (see Du et al 1996 for methods). Similarly, in a single choice test, significantly more oriented towards the treated plant (44%) than responded to the control (20%) (Birkett et al 1999). The volatiles released from plants that had previously been treated with the signal compound were also investigated and shown to contain heightened amounts of various compounds; for example, 48 h after treatment, β-ocimene was being released at twice the level produced by the control, and continued to be released at almost three times the original level in the following 144–192 h. This enhancement contrasted with the known signal methyl jasmonate, which resulted in a doubling of ocimene release in the 0–48 h period, but with no significant differences in the later periods.

Studies on phytopheromones have so far involved effects over relatively long periods which would, of course, be of great value in crop protection; thus, the promoter sequences and genes associated with responses to the recently discovered signal compound are now being investigated in collaboration with the IACR site at Long Ashton. However, other studies using new techniques (Agelopoulos & Pickett 1998) are allowing investigation of the rapid release of volatiles associated with the so-called oxidative burst caused by initial plant damage. Here, model studies using the western flower thrips, *Frankliniella occidentalis* (Thysanoptera), which damages a wide range of crops including ornamental plants such as *Chrysanthemum* spp. (Asteraceae), have found that the insect possesses specific cells for molecules typical of this oxidative burst, although considerably more work must be done before the biological significance of these responses can be determined.

Species-specific induction of plant volatiles by aphids

In the paper by Wadhams et al (1999, this volume), it is seen that the pea aphid, *A. pisum*, feeding on the bean *Vicia faba*, causes the plant to produce a range of compounds in addition to those normally produced on mechanical damage. However, when *A. fabae* or the vetch aphid, *Megoura viciae*, feed on the same plant species, a different range of volatiles is released (Du et al 1998). GC-EAG studies on the volatiles from plants damaged by aphids showed a number of active peaks, subsequently identified as being associated with (Z)-3-hexen-1-ol, (Z)-3-hexen-1-yl acetate, linalool, (E)-β-ocimene, 6-methyl-5-hepten-2-one and (E)-β-farnesene (Fig. 1). The behavioural activity of some of these compounds is described by Wadhams et al (1999, this volume), and the role of the methylheptenone has been described in a number of contexts (Quiroz et al 1997 and references therein, Du et al 1998). Clearly, this compound could be an isoprenoid produced by oxidation. We have also found the same compound, together with 6-methyl-5-hepten-2-ol and tridecanone, in the volatiles from wheat seedlings upon which *R. padi* was feeding. Various behavioural roles for these compounds have been found, as seen in Tables 1 and 2. Another compound which has been detected in these studies is geranic acid, which is formally an oxidation product of geraniol (Scheme 3), but a clear behavioural role for the compound has not yet been established. Biochemical work, particularly by Boland's group (e.g. Donath & Boland 1994), on the oxidative generation of homoterpenes such as 4,8-dimethyl-1,3,7-nonatriene (Fig. 2) is highly advanced, and these are seen to be important stress compounds and potentially useful

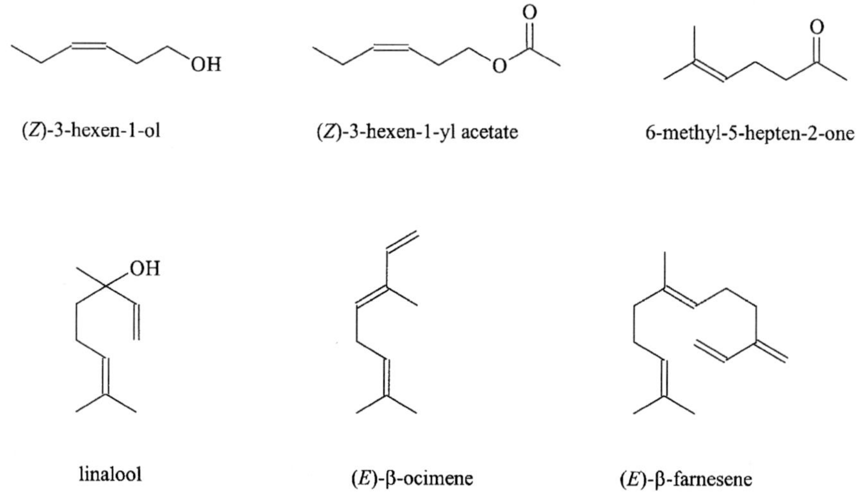

FIG. 1.   Structures from text.

semiochemicals for higher trophic level interactions. The work on aphid chemical ecology described here opens up other compounds for similar investigations.

Release of plant stress compounds from intact plants

Work in Africa on stem borer control, in collaboration with the International Centre of Insect Physiology and Ecology in Kenya, has identified a number of grasses (Poaceae) that can act as trap crops for stem borers attacking subsistence crops of maize and sorghum. Two species of grasses have now been used with great success in large-scale field trials and commercial on-farm trials: Napier grass, *Pennisetum purpureum* (a millet) and Sudan grass, *Sorghum sudanense*. Planting maize

TABLE 1   Response of apterous *Rhopalosiphum padi* in the olfactometer to a mixture comprising 6-methyl-5-hepten-2-one, 6-methyl-5-hepten-2-ol and 2-tridecanone (4:1:1, 0.01 µg)

Stimulus	Average time spent in each arm (min)	P
Mixture	2.11 ± 0.44	
Hexane	3.97 ± 0.48	0.005

TABLE 2   Response of apterous *Rhopalosiphum padi* in the olfactometer to undamaged wheat seedlings, with or without a mixture comprising 6-methyl-5-hepten-2-one, 6-methyl-5-hepten-2-ol and 2-tridecanone (4:1:1, 0.01 μg)

Stimulus	Average time spent in each arm (min)	P
Wheat seedling	5.22 ± 0.23	
Blank	2.49 ± 0.11	0.0013
Wheat seedling + mixture	2.16 ± 0.09	
Hexane	5.15 ± 0.85	0.0015

in 50 m squares surrounded by a trap crop, for example two rows of *P. purpureum*, gave a highly significant reduction of stem borers in the maize crop. In addition, the *P. purpureum*, although preferred to maize for oviposition, allowed only about one-quarter of the larvae to develop through to adulthood, thereby offering other ways of reducing the population, although this could interfere with the development of beneficial parasitoids. Similar trap cropping with *S. sudanense* resulted in reductions of stem borers in the protected maize to at least one third of the normal population, with populations in the trap crop being three to six times greater. Plants which are consistently not attacked by stem borers were tested as intercrops with maize, in the hope that they might repel stem borers away from the main crop. One of the most effective was found to be molasses grass, *Melinis minutiflora*, which, in a one-to-one intercropping with maize, resulted in less than one fifth of the percentage of plants damaged, compared to maize grown as a monocrop. This effect persisted even when the *M. minutiflora* was intercropped in a ratio of 1:3 with maize (Z. R. Khan, unpublished data).

Using coupled GC-EAG, the volatile semiochemicals responsible for attraction of stem borers such as *Chilo partellus* (Pyralidae) and *Busseola fusca* (Noctuidae) were

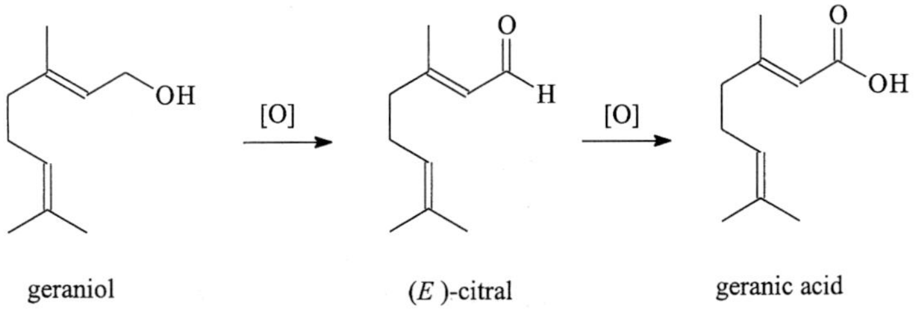

Scheme 3.   Relationship between geranic acid and geraniol.

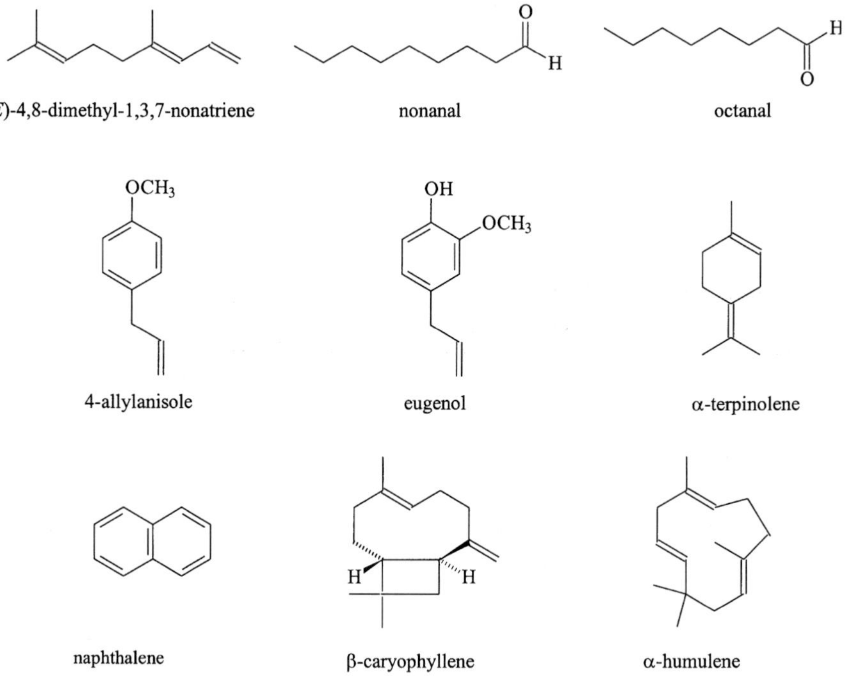

(E)-4,8-dimethyl-1,3,7-nonatriene   nonanal   octanal

4-allylanisole   eugenol   α-terpinolene

naphthalene   β-caryophyllene   α-humulene

FIG. 2.   Structures from text.

identified. These included nonanal, octanal, 4-allylanisole, eugenol, linalool and, surprisingly, naphthalene (Figs 1 & 2), all of which were confirmed to be highly attractive in the olfactometer bioassay (Z. R. Khan, unpublished data). On examining volatiles released by *M. minutiflora*, similar compounds were found which are components of many grasses, but additional compounds were found which were candidates for repellency against stem borers. Confirmed as such in the olfactometer were (E)-β-ocimene, α-terpinolene, β-caryophyllene, humulene and (E)-4,8-dimethyl-1,3,7-nonatriene (Figs 1 & 2).

As the multiple-cropping field trials developed, it was noticed that where stem borers were found in maize intercropped with *M. minutiflora*, parasitism was considerably increased, presumably as a result of enhanced foraging by the parasitoids. Since the nonatriene had been encountered previously as a foraging stimulus for parasitoids, it was examined in an olfactometer assay with the stem borer parasitoid *Cotesia sesamiae* (Braconidae) and found to have just such a role, accounting for almost all the attractant activity of the *M. minutiflora* volatiles (Khan et al 1997). Thus, it appears that intact plants can release compounds typical of stress or insect damage and which can, even under these conditions, cause attraction of parasitoids. For this strategy to be effective, it is important to be able to identify the compounds to which parasitoids respond innately, or which are important in parasitoid learning.

The work in Africa will now expand beyond Kenya and is set for success, not just in terms of crop protection, but also because of the value that the trap and intercrop plants have as cattle forage. Also, some intercropping systems have been shown substantially to reduce attack by witchweed, *Striga hermonthica* (Scrophulariaceae), and attempts are being made to exploit this aspect (Khan et al 1999).

Conclusions

Although considerably more progress must be made in determining whether naturally released signals can cause other plants to activate defences prior to attack, it is evident that insects can be used as a means of identifying compounds that may themselves be valuable signals for 'switching on' plant genes. Once the necessary promoter systems are identified, these signals could then be used to 'switch on' a range of genetically based effects, beyond crop protection and into nutritional traits and other plant or plant product attributes. Furthermore, the discovery that some plants release useful semiochemicals, in terms of increasing foraging by beneficials, even before damage, demonstrates the potential of an approach where plant signals could be used to cause intact plants to generate semiochemicals that might then help against pest attack. Extension of such studies may also provide explanations of why plants perform differently when grown with other plants, whether of different species or cultivars or as monocultures. These studies are helping to develop new strategies for controlling insect pests, and are also providing an insight into fundamental evolutionary aspects of insect–plant interactions.

Acknowledgements

IACR-Rothamsted receives grant-aided support from the Biotechnology and Biological Sciences Research Council, UK. This work was in part supported by the Ministry of Agriculture, Fisheries and Food, UK. The stem borer work was funded by the Gatsby Charitable Foundation.

References

Agelopoulos NG, Pickett JA 1998 Headspace analysis in chemical ecology: effects of different sampling methods on ratios of volatile compounds present in headspace samples. J Chem Ecol 24:1161–1172

Baldwin IT 1998 Jasmonate-induced responses are costly but benefit plants under attack in native populations. Proc Natl Acad Sci USA 95:8113–8118

Birkett MA, Campbell CAM, Guerrieri E et al 1999 Airborne plant signal identified. In preparation

Delledonne M, Xia Y, Dixon R A, Lamb C 1998 Nitric oxide functions as a signal in plant disease resistance. Nature 394:585–588

Donath J, Boland W 1994 Biosynthesis of acyclic monoterpenes in higher plants parallels steroid hormone metabolism. J Plant Physiol 143:473–478

Doughty K J, Kiddle G A, Pye B J, Wallsgrove R M, Pickett J A 1995 Selective induction of glucosinolates in oilseed rape leaves by methyl jasmonate. Phytochemistry 38:347–350

Du Y-J, Poppy G M, Powell W 1996 Relative importance of semiochemicals from the first and second trophic level in host foraging behavior of *Aphidius ervi*. J Chem Ecol 22:1591–1605

Du Y-J, Poppy G M, Powell W, Pickett J A, Wadhams L J, Woodcock C M 1998 Identification of semiochemicals released during aphid feeding that attract parasitoid *Aphidius ervi*. J Chem Ecol 24:1355–1368

Farmer E E, Ryan C A 1990 Interplant communication: airborne methyl jasmonate induces synthesis of proteinase inhibitors in plant leaves. Proc Natl Acad Sci USA 87:7713–7716

Hardie J, Isaacs R, Pickett J A, Wadhams L J, Woodcock C M 1994 Methyl salicylate and (−)-(1R,5S)-myrtenal are plant-derived repellents for black bean aphid, *Aphis fabae* Scop. (Homoptera: Aphididae). J Chem Ecol 20:2847–2855

Khan Z R, Ampong-Nyarko K, Chiliswa P et al 1997 Intercropping increases parasitism of pests. Nature 388:631–632

Khan Z R, Pickett J A, Hassanali A 1999 Management of the *Striga* weed in maize-based cropping system. Insect Sci Appl, in press

Nottingham S F, Hardie J, Dawson G W et al 1991 Behavioral and electrophysiological responses of aphids to host and nonhost plant volatiles. J Chem Ecol 17:1231–1242

Pettersson J, Pickett J A, Pye B J et al 1994 Winter host component reduces colonization by bird-cherry-oat aphid, *Rhopalosiphum padi* (L.) (Homoptera, Aphididae), and other aphids in cereal fields. J Chem Ecol 20:2565–2574

Pickett J A, Wadhams L J, Woodcock C M 1997 Developing sustainable pest control from chemical ecology. Agric Ecosyst Environ 64:149–156

Quiroz A, Pettersson J, Pickett J A, Wadhams L J, Niemeyer H M 1997 Semiochemicals mediating spacing behavior of bird cherry-oat aphid, *Rhopalosiphum padi*, feeding on cereals. J Chem Ecol 23:2599–2607

Shulaev V, Silverman P, Raskin I 1997 Airborne signalling by methyl salicylate in plant pathogen resistance. Nature 385:718–721

Visser J H 1986 Host odor perception in phytophagous insects. Annu Rev Entomol 31:121–144

Wadhams L J, Birkett M A, Powell W, Woodcock C M 1999 Aphids, their predators and parasitoids. In: Insect–plant interactions and induced plant defence. Wiley, Chichester (Novartis Found Symp 223) p 60–73

Ward E R, Uknes S J, Williams S C et al 1991 Coordinate gene activity in response to agents that induce systemic acquired resistance. Plant Cell 3:1085–1094

DISCUSSION

Vet: I was wondering whether you now know more about the effects of *M. minutiflora* on the stem borer, *C. partellus*, because when you published that paper in *Nature*, there was only a description that volatiles extracted by hydrodistillation of the plant reduced oviposition of the stem borer (Khan et al 1997). How much do you now know of real odours being involved in flight behaviour or settling, for example?

Pickett: The behavioural work is inadequate in formal terms. We used a test which does allow some contact with the substrate. The behavioural work has had to be done in Africa, under sometimes quite difficult circumstances, and cannot deal yet with some obvious questions. We have concentrated on interfering with the precolonization behaviour in the field, which inevitably involves principally the volatile semiochemicals.

Karban: You mentioned releases of methyl salicylate in field trials. Can you elaborate on how you did that?

Pickett: We quoted the release rates in milligrams per $10\,m \times 10\,m$ plot: the lowest release rate would equate to something like $5\,g/Ha$. This went out as emulsifiable concentrate with a vegetable oil to slow it down, or in sachets. We mostly use electrostatic spraying systems in this work.

Gierl: Complementary to the strategy to isolate repellents, could one also use the attractant as a starting point for modifications? Could you take a tight binder to make the insects blind for everything?

Pickett: So you would seek to habituate the insects to the attractants.

Gierl: Yes; could you block all the cells with a super attractant?

Pickett: That is an interesting idea.

Poppy: We had a student looking at *Plutella* sp. sex pheromone, in which she developed analogues to block the response to the sex pheromone. She basically swamped the environment with this analogue of the pheromone and then tried to get the *Plutella* to respond to a sex pheromone and it couldn't.

Pickett: The cyclopropene analogues involved are very stable, contrary to what might be expected. They work as good analogues of monoenes. In fact, sterculic acid is an inhibitor of lipid metabolism essential in membrane biochemistry and is, as a consequence, an extremely poisonous natural product. To inundate with the attractants is an interesting prospect, but I am doubtful that this could be achieved practically. However, if molecular geneticists such as Thomas Mitchell-Olds produce plants at both ends of the attractancy spectrum, this is very useful, because a push–pull strategy requires both aspects.

Schultz: I'm curious about the ecological implications of your methyl salicylate results. Do you know which, if any, of the aphids you study transmit plant diseases?

Pickett: In the summer there's a low population of *Rhopalosiphum padi*, which is the main disease vector, and the aphids causing mainly direct damage only include *Sitobion avenae*.

Schultz: Methyl salicylate could be taken as an indication of pathogenic infection; it could also be taken as an indication of drought stress, assuming that its emission is tied to its production, which goes up under both of those conditions. But I didn't really see a clear pattern there, with respect to what kinds of insects seem to be paying attention to it.

Pickett: If we knew how all insects respond behaviourally, then we could make some sense of it. We do know this for some of the insect species mentioned.

Wadhams: Almost all the phytophagous insects that we've looked at respond to methyl salicylate at the behavioural level.

Turlings: On your trap crop, the survival of the larvae was much lower: how did you measure this and how do you account for it?

Pickett: It was basically done on larval counts in the crop, by destroying plants to measure the level sequentially through the growth period. All the analyses were designed by statisticians: plants were taken out, slit down the middle and the larvae counted, throughout the life cycle. We think the reduced survival in Napier grass is largely associated with a mechanical defence: the plant produces a latex gum which engulfs the small larvae.

Gierl: What is the optimal distance of the trap plants from the main crop?

Pickett: We have precise views on this. We would like a 50 m × 50 m square of maize, with two rows of Napier grass a metre away from the crop. We were in a farmer's field the other day and she had grown the Napier grass closer to the maize, but this was still working satisfactorily.

Wadhams: Surprisingly, there didn't seem to be any higher levels of the stem borer infestation in the maize right up against the Napier grass trap crop. We've got some figures that if you use a 10% trap with your crop, you see a 16–20% gain in yield.

Pickett: With *Melinis minutiflora* we have done a lot of trials looking at spacing, and we've been surprised to see effects up to 1 row to 20 of maize. But we recommend 1 to 3 of maize because this gives the best overall effect.

Gershenzon: Is there a correlation between methyl salicylate emission and endogenous levels of salicylic acid? In the *nahG* plants, for instance, is there an interconnected pool? It would be very interesting if there was a volatile cue which allowed insects to estimate the status of the plant in terms of pathogen infestation.

Pickett: We have not studied this, but it needs to be done. We investigated a system for methylating salicylate, and spent considerable effort working with *Gaultheria* species without success. We did a great deal of feeding and labelling studies without getting anywhere.

Dicke: Lima bean plants emit relatively large amounts of methyl salicylate in response to spider mite infestation. The production of volatiles can be induced in lima bean with just jasmonic acid. Wilhelm Boland showed that jasmonic acid is induced by spider mite feeding. If salicylic acid were also to be induced, then this would block the jasmonate effect. Therefore, it seems that there is not a strong correlation between induction of salicylic acid induction and methyl salicylate emission.

Gershenzon: But suppose that methylation was just a way to dispose of salicylic acid. Then, there would be a direct connection between the pools of salicylate and methyl salicylate. The *cis*-jasmone story is also interesting in this respect, since it is believed that *cis*-jasmone is a volatile metabolite of jasmonic acid. Volatiles whose release is well-correlated with changes in the amount of endogenous signal molecules may tell us (and foraging insects) a lot about what's going on in the plant. Maybe lima bean is a good system in which to approach these problems.

Boland: Looking at this the other way round, coming back to plant–plant communication, what is the effect of the volatiles from the intercrop plants on the physiological status of the crop plants?

Pickett: That's a good question, and we have resolved to look at that. One of the intercrops which I didn't show, silver leaf (a legume, *Desmodium uncinatum*), is potent at repressing witchweed (*Striga* sp.) colonization of the maize. So there are very useful allelopathic effects.

Boland: Does it also turn on the defensive chemistry in the useful plant?

Pickett: We must investigate this as a priority case.

Firn: What about the long-term strategy of this as a means of control? Can you use your knowledge of the system to make it longer-lasting?

Pickett: We have thought about this a lot, because we're now embarking upon trying to expand the system from Kenya into Ethiopia, Tanzania and Uganda. There is a very rapid farmer take-up in Kenya which is only just under control. We are concerned that the strategy might, in becoming more widespread, be diluted, with only parts being applied instead of the full push–pull system. We are producing a manual together with the extension services to give the farmers the requirements for sustainable success. The seed will also be monitored by our African colleagues for the active compounds and periodically some electrophysiological and behavioural work will be done to see whether there is any genetic drift during seed production. The extension service will also maintain its own field trials to make sure that the population does not suddenly become resistant. We really need long-term funding to ensure this can continue.

Reference

Khan ZR, Ampong-Nyarko K, Chiliswa P et al 1997 Intercropping increases parasitism of pests. Nature 388:631–632

Final discussion

Gershenzon: I think this is a good time to ask a more general question about some of the methods that we've seen here. Perhaps someone could speculate on methodological advances, particularly in the analytical area, which might impact this field over the next few years. So much of the information that we have now has only recently been obtainable. There have been fantastic advances in volatile collection methods recently, such as the development of solid phase microextraction (SPME) fibres and the system employed in Jim Tumlinson's group. I'm curious about what might be on the horizon that will lead us to further improvements in this area.

Boland: There is a new method, called photoacoustic spectroscopy, in which one can induce certain molecules by laser energy and measure the decay of energy by acoustic spectroscopy. This is being developed at several universities. It gives a very high time resolution. If one follows a molecule like methane, ethylene or butane, this can be time-resolved on a second scale or lower. Responses to damage could be measured in real time with this technique. These types of on-line methods which are highly sensitive should be a goal for the future.

Beale: Isn't ethylene the easiest compound to detect with this photoacoustic spectroscopy?

Boland: It can actually be tuned to a variety of compounds.

Beale: Methyl jasmonate is the obvious one that people have tried, but there's so much interference from the carbon dioxide that it is just not sensitive enough.

Pickett: The problem we face, is that we often have a flat baseline on the chromatographic (e.g. GC) trace where we know there is activity. We are several orders of magnitude less sensitive in our chemical analysis compared with what the insect can respond to. In these cases we try to link the GC electrophysiological preparation to another system — e.g. a behavioural preparation with the honeybee proboscis extension or wasp wing fanning. This gives more assurance that a particular region of the chromatogram is associated with biological activity. Collection from large numbers of insects does not necessarily give more information. In GC there is a particular problem of large inactive peaks close to the active region. In terms of identification, we are struggling with the current range of sensitivity. We are synthesizing tentatively proposed structures with less and less information. Do you find that also, Jim Tumlinson?

Tumlinson: Yes.

Vet: Working with these individual compounds is problematic. The interpretation of the sensory input is so complex. I don't think we should hunt for a particular volatile by itself, but instead we should get more insight into the whole process of information processing. How is the sensory input translated into behaviour? For this, the behavioural assays remain essential.

Pickett: We could do a lot more at the behavioural level. We can observe many different types of behaviour. Automatic recording systems have improved. If only oviposition behaviour is employed, much would be missed.

Vet: It is essential to look at what compounds do behaviourally. If you see something specific, then you can key in on linking a particular compound with a particular behaviour.

Tumlinson: I think the point that Wilhelm Boland and others are making is that we need to be able to detect the compounds that are being released by the plants at exactly the same time that the insects are detecting them. This is the only way you really know what the insect is responding to. We've tried doing this by splitting air streams, but this is very tricky.

Pickett: If the semiochemicals in question are produced by plants, it would be most appropriate to use the plant itself. With plant molecular biology, I think we are going to have the opportunity to create plants with and without expression of plant-derived semiochemical genes for these studies.

Boland: Another methodology one should at least mention it is the so-called electronic nose. These are chips that can measure on-line components that they are programmed for. I doubt whether the sensitivity will be sufficient for our purposes.

Pickett: I am strongly critical of the electronic nose: I think the claims made for these devices are unsubstantiated. In the systems that are currently being developed, there is no molecular recognition involved — only the recognition of physical interactions which relate at best to a homologous series of chemical structures, rather than as sensory neurons to the individual compounds.

Boland: However, this method has one advantage, in that it is on-line with a very short time resolution. If we could train such an electronic nose to a specific compound or a set of compounds we would at least be able to follow this in real time. Time resolution could be a coding method for volatile messages.

Pickett: The pharmaceutical industry has created molecular recognition systems in the form of antibodies: moderately large organic molecules can be made antigenic by linking to antigenic proteins. But this approach does not well accommodate small lipophilic molecules. This is why at Rothamsted we have a molecular biology programme on molecular recognition in olfaction.

Beale: There are two problems here. You are interested in compounds you don't yet know about — that is, finding new active molecules. Secondly, you have the problem of analysing very low amounts of compounds that you have already identified.

Tumlinson: No, I think we're just as interested in structures that we do know, but even where we know what the structures are, we still don't know in a short period exactly what proportions of these compounds, or what concentrations are being released. The time resolution is important.

Beale: Surely, if it's a compound you know, can you not use the insect receptor which is the specific binding site for that molecule to build some sort of biosensor?

Tumlinson: This has been done for insect pheromones, but it is still rather complex and difficult.

Wadhams: It is a feasible proposition, but it will be difficult to do with the systems that are currently available.

Poppy: Getting back to the behaviour end of things, even if we have a receptor which is sensitive enough, this is still not answering the question of whether central integration of that message results in a motor response. The chances of being able to do behavioural experiments on-line in the field are small, but this would be a challenge. Biologists and behaviourists are often considered to be the poor cousins of chemical ecologists. We might not be heavily method-driven, but it's the interpretation of what we find in bioassays and designing rigid bioassays that is our skill. Being a good behaviourist is like being a good mass spectroscopist who is good at interpreting spectra rather than just searching a library. If we are going to pursue these things further, it's all very well keeping them technology-driven, but the skilled behaviourists are still needed. We must always remember that it is only when you discover the biological relevance of your chemical signal that the whole thing means anything and is not just noise.

Tumlinson: I agree.

Poppy: I think in the next 10 years the work of people like Laurent and Hildebrandt in terms of resolving the central integration within the brain will prompt some major leaps forwards.

Dicke: But this does not resolve the problem of making the complex mixtures that are used by carnivorous arthropods. With the methods just discussed you can see what the insect does in the brain with all the information. If you want to do that with the complex blends that we are dealing with, I am not sure that in the next decade we will make a big leap forward. The final answers should be given at the behavioural level, i.e. elucidating the mixture composition that is important to the animal in discriminating it from another mixture. To do this we need to be able to make complex blends similar to what plants produce and that is a real challenge.

Poppy: The on-line thing is interesting, because at the behavioural level we've always tried to explain variability in terms of barometric pressure, or the parasitoids having an off day or whatever — we've never been able to collect the volatiles from the same plants that we are doing the behaviour on. These new systems, in enabling us to collect for just a few minutes, do give us the ability to collect from a plant, do a bioassay on that plant and then collect from that plant again, in order to tie down whether the variability is from the parasitoid or the plant.

Pickett: We have arrived at the stage where we have some fairly clear take-home messages. These are complicated and must be viewed in the light of the entire discussion presented here. I always measure the success of each day by what I learn. In this respect this meeting has been superb: I have learnt an immense amount. I believe we have had an opportunity to hear some very different views. Although we seem overall to have been more-or-less in agreement, in reality we have had many different views expressed. This has been in a very positive atmosphere, which has facilitated useful discussion of these views. I congratulate you all for managing to achieve this, and I extend my thanks to all the speakers, the discussants and to the Novartis Foundation.

Index of contributors

Non-participating co-authors are indicated by asterisks. Entries in bold indicate papers; other entries refer to discussion contributions.

Subject index

A

Abies grandies see grand fir
abietadiene synthase 134, 138
acifluorofen 188
Aculops lycopersici 44
Acyrthosiphon pisum 41, 60–63, 256, 257
   Aphidius ervi and 34–35, 49, 60–61
aggregation pheromones 129
agriculture 1, 17, 225
alfalfa 189
allelopathy 129
allene oxide cyclase 198
allene oxide synthase 198–201
Alternaria brassicae 36
Amblyseius andersoni 46
Amblyseius finlandicus 46
amino acid conjugates 114–117
analogues, use in crop protection 221
angiosperms 136, 139–140, 165
anthranilic acid 182
Aphidius eadyi 65
Aphidius ervi 9, 41, 60–66, 256
   Acyrthosiphon pisum and 34–35, 49,
     60–61
   sex pheromones and 65, 67–68
   volatile release response 16, 34–35
aphids 60–73, 254–255, 257–258
   alarm pheromones 68–69
   hypersensitive response 71–72
   life cycle 64, 68
   mutants 69–70
   oral secretions 54, 72
   parasitization 16–17
   pathogenesis-related proteins 93
   pesticide resistance 69–70
   sex pheromones 62, 63–65, 67–68, 71
   virus infection 72–73
Aphis fabae 61, 73, 254–255, 257
apical meristem 90
apple 46–47, 226
apple rust mites 226
Arabidopsis 170, 179–181, 211

Arabidopsis thaliana 239–252
   candidate gene analysis in 242–243
   enzyme activity in 245–246, 249
   genetic variation in 248–252
   glucosinolates and 250
   glycolysis in 245–246, 249
   α-linolenic acid in 192
   mutants 250–251
   Plutella xylostella and 246, 248, 250, 251
   quantitative trait loci (QTL) mapping in
     242–243
   salicylic acid (SA) and 200
   signalling pathways 195
Arabis 10, 252
arachidonic acid 127–128
atrazine 150

B

bacterial artificial chromosomes 239
bark beetle 41
barley 155–156
barley yellow dwarf virus 72
beet armyworm 54–55, 96, 100, 101, 163
behavioural analysis 267–269
benzothiadiazole *see* benzo(1,2,3)thiadiazole-
   7-carbothioic acid S-methyl ester (BTH)
benzo(1,2,3)thiadiazole-7-carbothioic acid S-
   methyl ester (BTH) 184, 208–210, 227–
   228, 237
   Bion® and 209
   fungicide and 212
   pathogenesis-related genes and 230
   secondary metabolites and 217
   systemic acquired resistance (SAR) and
     208–209
   vaccination of crops and 224, 229–230,
     236
bergamotene 80, 89
biological control 71, 96, 104, 158, 253
   aphids 65–66, 71
   see also crop protection
Bion® 209, 211, 218–219, 227–228

272